Beyond the Quantum Paradox

To all my dear friends and, the dearest one, my wife ZILMA

Beyond the Quantum Paradox

Probability riddles . . . Quantum riddles . . . Other riddles . . .

Lazar Mayants

Taylor & Francis
Publishers since 1798

UK Taylor & Francis Ltd, 4 John St., London WC1N 2ET

USA Taylor & Francis Inc., 1900 Frost Road, Suite 101, Bristol PA 19007

British Library Cataloguing in Publication Data

A catalogue record for this book is available from the British Library

ISBN 0-7484-0206-3 cloth
       0-7484-0207-1 paper

Cover design by Linda Wade

Typeset in Great Britain by Keyword Publishing Services
Printed in Great Britain by Burgess Science Press, Basingstoke
on paper which has a specified pH value on final paper
manufacture of not less than 7.5 and is therefore 'acid free'.

Contents

Preface

We think in generalities, but we live in detail
Alfred N. Whitehead

One is not bound to believe that all the water is deep that is muddy
Thomas Fuller

Have you ever heard of the chicken–egg paradox? It declares that it is
impossible to decide whether the chicken or the egg comes first. Or have
you heard of the paradox about the cat which, while in a closed container,
is neither alive nor dead—according to those physicists who accept conventional quantum physics? The first paradox seems insoluble in principle. The
second is regarded by many as one that cannot be settled in any reasonable
way.

Perhaps you have also heard of another famous enigma in physics, the
diffraction of particles, which demonstrates that particles have not only
corpuscular properties but also wave properties. The question of how one
entity can move as a particle and yet, simultaneously, propagate as a wave,
has been debated by scientists for many decades. Perhaps you have heard
of some experiments, both imaginary and real, intended to check whether
the supposed simultaneity of the corpuscular and wave properties of a
particle is real, and whether quantum physics is truly compatible with our
ideas of reality. The results of these experiments, however, are controversial
and even paradoxical, and have also led to endless debate. And what about
quantum physics itself? Do you have any idea of what it actually is? It is
hardly any wonder if you do not, for scientists themselves seem not to
understand it. Prominent physicists and philosophers have quarrelled frequently over the meaning of quantum physics and still there is no agreement. One highly respected physicist has even stated that quantum physics
cannot be understood, that one must simply get used to it, a contention that
has reconciled most physicists to the bizarre situation that exists in their
science.

The only aspect of quantum physics with which all scientists agree is that
it is connected with probability. But what, then, is probability? You perhaps
know that it is somehow related to statistics. But why? And how?

You have, furthermore, certainly heard of photons, of particles of light.
But what, again, are they? What is their nature? There are many other
puzzling things, seemingly unrelated to science, which you have also heard
of, among them the recent cataclysmic failure of communist ideas. Is there

any principle which prevented them from realization? Finally, let us ask two more questions: who are 'you' whom I am addressing?, and who am 'I'?

Your present reaction is probably one of surprise, and you may well be ready to protest: the chicken–egg paradox, the mysterious cat, diffraction of particles, quantum physics, photons, communism's failure, and even our respective existence as people—can there be any possible connection between all of these matters?

The answer is in the affirmative and is contained in this book, in which I shall endeavour to present a clear exposition, entirely without mathematical formulas, of a new approach to the treatment of the subjects involved. The material is of great significance but the presentation is in the popular scientific vein, and as such is intended for the curious layman as well as for students of science at all levels. The new approach settles the aforementioned seemingly insoluble riddles, provides a full understanding of quantum physics and banishes its hitherto unresolvable paradoxes, explains the nature of photons, and even indicates why communist ideas have failed. It also answers our earlier, seemingly strange, questioning of who you are and who I am.

Let me explain why I have decided to write this book and what has made me confident in challenging the situation in conventional quantum mechanics—aspects that may intrigue my readers. In connection with the decision to write, it is interesting that when I told one of my colleagues that I was preparing a book on quantum physics with no mathematics in it, he exclaimed: 'But that is an incredibly hard task!' He certainly would be quite right if I tried to popularize conventional quantum mechanics. Such a task is not merely hard—it cannot be accomplished in principle, I believe, at least for the aforementioned reason that no common agreement exists on what quantum mechanics is. What I am doing, however, is introducing *quantum physics* as a *whole science*, that is, one having both theoretical and experimental parts, and this makes a difference, because the essentials of any whole science can be set forth and reasonably explained without resorting to any mathematics.

This book is based on the author's own works published within the last few decades (1964–1991). I have never before written any popular essays, except for a few articles for some Soviet encyclopaedias, and I should not have written this book either were it not for a challenge by a friend of mine, not an expert in natural sciences, who was very interested in understanding the oddities of the conventional quantum mechanics he had read about in some 'popular books' knowing also that I had settled them in quite a realistic way. His challenge was strongly supported by my wife, who knew all about my work, and this convinced me to undertake the task. A further spur was my longing to try to break the vicious circle existing in physics education. The students learn from their teachers how to perform quantum mechanical calculations, but they do not understand what quantum mechanics is. Then they become teachers themselves and do the same thing to a new generation of students. This process of inadequate training has been going on for quite a time all over the world. I hoped that this book would facilitate a change in that unfortunate situation.

My confidence in my right to challenge can be understood from my long trek through science, along paths full of obstacles, in search of the solution to the vital problems which were permanently exciting and bothering many contemporary physicists and philosophers.

As a high-school student in Russia, I was deeply interested in both mathematics and organic chemistry. Before entering a university, I worked at a chemical institute synthesizing some rather complex reagents. I began to wonder why a chemical reaction proceeded, not in any of many possible directions, but often just in one. I soon realized that to try to solve this problem I would have to study physics more profoundly. Therefore when I attended Moscow State University, I specialized in physical chemistry and became acquainted with quantum mechanics.

After World War II, having a Ph.D. degree in chemistry, I was able to continue my research in theoretical molecular spectroscopy at the Lebedev Institute for Physics, where I was awarded a D.Sc. degree in physics and mathematics. To solve some important problems, I had to invent some new computational methods in linear algebra. After a seven year compulsory interval, during which I taught general and theoretical physics at universities outside Moscow, I was invited back to an Institute of the Academy of Sciences of the USSR and continued my research in theoretical molecular spectroscopy.

I could not forget, however, my youthful desire to try to tackle the kinetics of chemical processes. At the end of 1962, I finally proposed a theory of the kinetics of chemical and physical processes which contained a criterion for telling whether or not a certain intramolecular rearrangement is feasible. My theory seemed to be in disagreement with the common views of quantum physics, and this led me to scrutinize its conventional foundations. I found them rather unsound and began exploring the problem more thoroughly.

At that time, I was able to disprove only the common belief that quantum mechanics is incompatible with reality, but I did not yet understand its real meaning and was not even sure that I ever would. It was a long and difficult search, but in 1966 I concluded that I had found the answer when the connection I had anticipated between quantum mechanical and classical probability distributions for an important particular case was confirmed by calculation. I now was able to construct quantum mechanics anew, proceeding from some realistic premises, but the premises themselves still required clarification, and I needed to dig even deeper.

Finally, one day in 1971, a new idea dawned on me—I came across the new principle that put everything in its place. It settled many paradoxes and issues immediately, and provided a sound basis for all probability related sciences. However, the area of its application was even wider, as the reader will soon discover. I started with an attempt to find out what was perhaps wrong with the conventional construction of quantum mechanics and ended with results whose scope happened to be much wider than physics alone.

The full statement of my approach was set forth in 1984 in the book *The Enigma of Probability and Physics*. To give some indication of the impact this approach had on certain leading physicists, I will allow myself to cite just three sentences by Professor Henry Margenau of Yale University. 'In this book [Mayants] presents a unique, extremely detailed, and embracive version of a subject that has suffered for a long time from numerous internal imperfections. His approach is new and original, the material covered features not only the foundations of the science of probability but also most of its applications, including statistical and quantum mechanics. The key methodological principle underlying the book is of extraordinary significance and deserves special attention.'

This methodological principle I shall discuss presently. I have tried to make this and all the other subjects in the book as accessible as possible to a wide range of readers without simplifying to the point of possible distortion. I trust that my readers will study it thoroughly and open-mindedly.

Lazar Mayants

Acknowledgements

I greatly appreciate the attitude towards my work shown by Arnold Silver, Professor of English at the University of Massachusetts in Amherst. He went through the manuscript thoughtfully, improving its style and enhancing thereby its clarity and intelligibility. But I am especially thankful to him for his steady interest in my ideas and their applications. Our conversations on this subject, in particular, were a major stimulus to undertake the task of writing this book.

My thanks go also to Professor William Mullin, of the University of Massachusetts Physics Department, who read the book in manuscript and provided me with useful comments on it.

I owe, finally, innumerable thanks to my wife, Zilma Mayants, a Professor of Literature and Aesthetics, who participated in the discussion of all parts of the work and was especially helpful with its methodological and philosophical aspects.

Chapter 1
Schrödinger's cat: what is it about?

Is the cat alive or dead?
The answer had read:
'Neither this, nor that'.

1.1 Preliminaries

Readers unfamiliar with the cat mentioned in the above heading and the epigraph should first acquaint themselves with it (see Section 1.2). This (i.e., Schrödinger's) cat involves quantum mechanics. The very sound of these mysterious words 'quantum mechanics' may make the less initiated tremble, as if the subject were something beyond the comprehension of ordinary people. And what about the initiates?

Most of them do not care about the meaning of this domain of physics—they simply know how to calculate the quantities they are interested in, and that is all they need. Many, however, have tried to comprehend what quantum mechanics is, but have not succeeded. They are divided in their opinions on how to understand it, and on what is its meaning. One of the indicators of this bizarre situation is the fact that, after many years of discussion, scientists (physicists, philosophers, and others) are still debating the problem of how to settle in particular, the so-called 'cat paradox' by Schrödinger, and indeed many other paradoxes related to quantum mechanics. This circumstance makes quantum mechanics look like a kind of religion rather than a genuine science. And this is not at all by accident.

Historically, quantum mechanics was founded on a then rather new branch of pure mathematics, and its original goal was the *ad hoc* explanation of some new experimental facts in optics (spectroscopy). Attempts to understand its physical meaning began somewhat later, without much success, and are still continuing. The only really important thing which had been understood correctly is that some mathematical quantities involved were very closely connected with probabilities. The mathematical origin of conventional quantum mechanics is reflected even in the names of some conferences dedicated to the discussion of its issues, as for instance, 'Mathematical Foundations of Quantum Theory'.

This saturation with mathematics seems to render hopeless any attempt to explain the essence of quantum mechanics to an ordinary audience, the more so because the scientists themselves do not have a common opinion on it; most of them agree with the current aphorism attributed to the well known Soviet physicist L. Landau: 'Quantum mechanics cannot be understood—one must simply get used to it'.

That is too bad for a science, but it occurred because the construction of

1

quantum mechanics began from its roof—its mathematics. Had it had been built from its basement—its essence—no issues of the kind under common discussion would have arisen at all, and there would be no serious difficulties in explaining its meaning to anybody wishing to understand it. Fortunately, such a new construction already exists, based on an appreciation of quantum physics as a whole science. The mathematics needed happened to be basically the same as in conventional quantum mechanics. However, it is supported by a full understanding of *why* it should be just the way it is.

It is precisely this new situation which allows me to explain in the following pages, at least qualitatively, the essence of quantum physics (as a whole science). At the same time, many current issues are thereby settled, and readers will be able to reassure themselves that there is no mystery at all in quantum physics. In this context, Schrödinger's cat will help us in our attempt to find the right way to the truth.

The interesting and important point is, and I stress it again before going on, that the new treatment of all the matters involved results, in the final analysis, from a very simple and at the same time very general key principle whose sphere of application is extremely wide (see Section 2.5).

1.2 The cat paradox

Schrödinger suggested in 1935 a thought experiment which can be presented in the following version.

A cat and a flask of poison are enclosed together in a hermetically sealed opaque container. If the flask is broken, the cat is killed by the poison. Breakage of the flask is triggered by the discharge of a Geiger (counter) placed behind one of two holes made in a screen which is irradiated at the front by a beam of electrons. It is supposed that the beam is so weak that only one electron passes through the screen during the whole experiment which can last for a long time.

Upon completing the experiment, the container is opened and the cat is found either alive or dead. But what can be said about the state of the cat during the experiment, before the container was opened? Was it dead or alive?

Realistically thinking people would suppose that if the electron passing the screen does not enter the Geiger-related hole, the flask will not break and the cat will be alive all the time; otherwise it will stay alive only until the moment the electron hits the Geiger, and be dead after that has happened. This seems clear and reasonable, and many readers of this book, especially the 'uninitiated', would agree, I believe, with such an answer, which would also satisfy, perhaps, everyone in the pre-quantum-mechanical era.

Now one of the initiates, a specialist in quantum physics, comes along and says: 'no, you are wrong, my friends; the situation is very different', and he starts to explain the quantum mechanical point of view, which is roughly as follows.

The state of the cat during the experiment is determined by the state of the oncoming electron. However, it is completely uncertain which of the two holes in the screen the electron has entered. There is nevertheless a definite probability for each of the holes that the electron passed through it, which coincides with the probability that the cat is alive or dead.

So far so good, and most reasonable people would subscribe to this account. Now, however, the formalism of quantum mechanics (its mathematical techniques) enters the game (readers should not be alarmed: no mathematical formulas will be used, as promised), and a special auxiliary function, the so-called wavefunction, is introduced, which describes the state of the electron and, hence, of the cat too. The function is the sum of two terms, each of which describes the state of the electron passing the respective hole, and it is the square of the term, not the term in itself, that gives the probability. The uncertainty of the state of the cat is thus given by the above function, and the probability that the cat is alive or dead is determined by the respective terms in it. The crucial point is that the function should be regarded as an indivisible whole—it cannot be split into those two terms. This means that the cat is neither alive nor dead during the experiment. If the conditions of the experiment are such that the probability of entering either hole is equal to 1/2, then the cat *during* the experiment should be *alive-dead* (or *dead-alive*, which is the same). It becomes either dead or alive only when the experimenter opens the container to find out what has happened to it, that is, when (in quantum mechanical terms) the measurement is made.

The above conclusion would seem paradoxical to realistic thinkers, the more so because the experimenter is free to postpone the examination of the cat's state for an indefinite time interval during which the cat must stay 'alive-dead' (neither alive nor dead)! Is this not absurd? The cat paradox prompted much debate which has not ended till now. Before going on, some consideration will be given to paradoxes in general, and to the part they are playing in the development of science in particular.

1.3 A little about paradoxes

A paradox generally is a situation exhibiting contradictory, incompatible aspects, or a statement that is essentially self-contradictory, although based on a seemingly valid deduction from acceptable assumptions. There are paradoxes that can be settled easily without any change in what is already known. They arise usually due to either wrong assumptions or some flaw in the logic of reasoning. Mathematics sometimes uses paradoxes for proving particular propositions. Roughly, if the supposition that the proposition to be proved is wrong leads, based on a valid deduction, to a paradox, then the proposition is true.

By way of example we consider here two historically famous paradoxes brought up by philosophers and logicians.

1.3.1 *Zeno paradox*

Zeno, a famous Greek philosopher, suggested the following paradox. Achilles, the mythological Greek hero, famous for the speed of his running, would never overtake a tortoise—one of the slowest creatures on Earth. His reasoning is as follows. Suppose Achilles tries to overtake the tortoise, both moving uniformly. By the time he reaches point 1, where the tortoise was originally, the latter has already moved further to point 2; by the time he reaches point 2, the tortoise has already moved to point 3; and so on, without any end. The endlessness of this process seemingly proves Zeno's assertion, which obviously contradicts our experience. The problem of when Achilles will overtake the tortoise is an easy algebraic one for high-school students. What, then, is wrong with Zeno's reasoning?

The flaw in his logic can be found easily. The distances between the adjacent points 1, 2, 3, . . . and, hence, the time intervals Achilles runs them through make a decreasing geometrical progression. The time needed to overtake the tortoise, which is the sum of the pertinent time intervals, is finite, even though the number of intervals is infinite. Zeno supposed that the infinity of the number of space intervals run through by the 'racers' entailed the infinity of the time Achilles needed to overtake the tortoise, and that was wrong. Incidentally, the process suggested by Zeno can be used for calculating that time. The result is the same as that obtained by solving the appropriate algebraic equation.

1.3.2 *Cretan paradox*

One Cretan has said that all the Cretans lie. The question is: is he himself lying or telling the truth? If he is telling the truth then, being a Cretan himself, he must be lying too. This means that all the Cretans are telling the truth. But then his assertion should be right, that is, all the Cretans, including himself, must be lying. And so on.

The inner contradiction related to this paradox seems not to have a logical solution. There is, however, a flaw in the reasoning, which tacitly presupposes that the only postulate contrary to 'all the Cretans are lying' is 'all the Cretans are telling the truth'. But there is another option, namely, 'not all the Cretans lie', and this settles the paradox. Indeed, now the answer to the question whether the Cretan mentioned first is lying or telling the truth is: he is lying, but not all the Cretans do so, and this answer does not lead to any inner contradictions.

There are, however, paradoxes that demand for their settlement new ideas, new principles. It should be emphasized at once that no paradoxes are inherent in nature itself. They may appear, however, when one tries to understand and explain the observed facts. The appearance of a seemingly insoluble paradox means, therefore, that something is lacking in our understanding of natural phenomena or social issues, or perhaps something is wrong in the way we are trying to get to know them.

Paradoxes of that kind are of the utmost significance, for when the principle needed is found, it not only resolves the particular paradox but also opens new prospects for further development of the pertinent science.

By way of example, recall the paradox that scientists came across when examining the velocity of light. That velocity was found to be one and the same, irrespective of the velocity of both the light-emitting source and the reference system in which the light velocity was measured. This fact contradicted Gallileo's well known relativity principle of classical mechanics that also was based on facts. The paradoxical situation was that the experimental facts about light contradicted the observations in conventional classical mechanics of those times.

Einstein had understood that to get rid of that paradox it was necessary to repudiate the conventional premise of those days, which asserted that space and time are independent of each other, for it was just that assumption that led to the paradox. He introduced a new physical principle which contended that space and time were inter-connected, forming a united four-dimensional space-time continuum. The paradox was thereby fully settled, and the far-reaching consequences of Einstein's epoch-making greatest discovery (published in 1905) are well known.

There are paradoxes, too, of another kind, which cannot be resolved without making use of proper new principles. Consider, for instance, the question: which comes first, the chicken or the egg? This question is usually considered insoluble. Indeed, if we assume that the chicken comes first, there must be an egg from which it was hatched; so the egg comes first, but then it must be preceded by a chicken which hatched it; and so on. The impression is that no solution can be found to this paradox. It is obvious that no new principle of natural sciences may be helpful in settling the chicken–egg paradox. This paradox is actually due to the inadequacy of our conventional way of thinking, and the new principle needed must, therefore, be related to our perception of reality.

It may sound rather strange, but the cat paradox is of the same kind. A number of quantum mechanical paradoxes as well as some others in classical statistical mechanics, and even in probability theory, are also of that kind, and the necessary principle discovered[1] by the author has settled them all. This principle will be set forth in more detail later. For its better understanding it would be useful to return now to Schrödinger's cat.

1.4 More about Schrödinger's cat

The fact that the cat paradox is still under discussion undoubtedly indicates that something is wrong with the conventional quantum mechanical treatment of the cat experiment itself. To try to find out where the flaw is, let us think the experiment through more thoroughly. We try, first of all, to separate its essential features from the nonessential ones.

It is obvious that no particular features of the cat sitting in the container, such as its sex, colour, or age, for instance, are essential to the outcome of the experiment. Nevertheless, it should be a *definite* cat that really exists, i.e. some *concrete* cat (a particular instance of a *concrete object*, see Section 2.2). The kind of live being chosen for this particular experiment, which involves killing, need not be a cat. This can be replaced by a mouse, for instance, or whatever else the experimenter may elect, even by a man, as was suggested by Wigner. The way the killing is achieved does not affect the outcome either: the poison can be replaced by any lethal device. The killing of a live being was chosen by Schrödinger, in the author's opinion, to set off the paradoxical features of the conventional interpretation of quantum mechanical formalism. Otherwise, instead of the cat-and-poison container, any device could be used which would indicate that the electron has entered just the hole behind which the Geiger is placed. It can be, for instance, a bulb switched on by the Geiger, or whatever else an experimenter can devise. So the cat serves merely as an indicator that a definite event, namely, an electron passing through a definite hole, has occurred.

Incidentally, Schrödinger's critical attitude to the conventional interpretation of quantum mechanics can be seen from the way he introduced his cat paradox. I am citing here only two sentences out of 20 one-column lines about it,[2] which show that he had actually ridiculed that interpretation. He starts with the words: 'One can construct also completely burlesque examples'. Then, after describing the experiment, he concludes: 'The psi-function of the entire system would express this by stating that the alive and dead cats (if I may say so) are in equal parts mixed or smeared' (translated from the German).

Now, about the part the electron plays in the experiment: it is obvious that no specific features of the electron, such as its speed, origin, etc., are essential. However, since only one electron passes the screen during the entire experiment, it should be a *definite*, really existing one, that is, some *concrete* electron. An electron is elected to cause the Geiger discharge, but any other charged particle, such as a proton or an α-particle, for instance, can be chosen to that end. What is more, since the cat is considered an indicator of the occurrence of a definite event, the electron can be regarded as a participator in that event. (In Schrödinger's original version[2] the relevant event is the decay of an atom of a radioactive sample, which triggers the discharge of the Geiger.)

After we have singled out the essential features of the cat experiment, it is easily seen that this experiment is a very particular case of a general one which may be depicted schematically as follows.

There is a device consisting of two parts, A and B (see Figure 1). In part A, only one of two mutually exclusive events, say a or b, can occur with a definite probability during the entire experiment. In part B, there is an indicator which reveals which of these events has occurred in reality. In the cat experiment, Part A is the screen with two holes, whose front is irradiated by a beam of electrons; the passing of the electron through one or other hole is the mutually exclusive events. Part B consists of the Geiger and the container holding a flask of poison and the cat (awaiting its fate!).

Extending the quantum mechanical reasoning concerning the cat experiment to the general case, one may believe that the state of the indicator in part B of the device is uncertain until the experimenter has looked at the reading. Imagine, for instance, a computer selecting, at random with equal probabilities, one of the two numbers 0 and 1, and keeping the recorded result in its memory. From the conventional quantum mechanical standpoint, the recorded result is uncertain (not merely unknown!) until somebody reveals it, which can in principle happen in an indefinite time. During all that time it is neither 0, nor 1 (it should be considered zero-unity or, which is the same, unity-zero). Is this not paradoxical?

One can invent many other examples of paradoxical conclusions of this kind, all involving probabilities.

Some readers may doubt the legitimacy of the above example. They may query what a computer has to do with quantum mechanics, or point out that quantum mechanics deals with microsystems, whereas a computer is a macrosystem.

The answer to such comments is simple. To say nothing of the cat, which is also a macrosystem, it should be remembered that the function describing a state of a system in quantum mechanics is an auxiliary mathematical quantity somehow related to probability. As I have shown elsewhere,[3] the formalism of quantum mechanics is specific not just for quantum mechanics exclusively—it belongs to probability theory in general. (It is precisely the fact that quantum mechanics involves the *probabilistic* treatment of physical

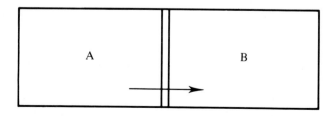

Figure 1.   *An event occurs in part A of the device and is registered in its part B.*

(micro) systems which determines the applicability of that formalism in quantum mechanics.) Therefore it should be valid also for the probabilistic treatment of macrosystems. (Note that mathematical methods specific for the probabilistic treatment of macrosystems are, as experiments demonstrate, inapplicable to microsystems.)

Thus our consideration of the cat paradox and similar ones makes us suspect that they are due to some flaw in the interpretation of the related quantum mechanical formalism. The key word in that interpretation is 'probability'. We should, therefore, give careful consideration to the meaning of that word.

Chapter 2
Probability: a new approach

2.1 What is probability?

The word probability is used widely both in everyday life and in many sciences. In everyday life the meaning of probability and the related words probably and probable seems self-evident to everyone who uses them. These words generally imply a shade of doubt as to whether or not a certain event will occur; they convey thus a subjective estimate of the likelihood that the event in question will take place.

The most important use of the word probability, however, is as a value between 0 and 1 assigned to an event to estimate the likelihood of its appearance. A systematic treatment of probability as a numerical quantity, that is, the emergence of probability theory, seems to go back to the 17th century. This theory deals mainly with 'random events', that is, events which are the result of chance alone. It was intended, in the first place, to predict by means of calculation a measure of the likelihood that the random event in question would occur. Incidentally, the very first applications of probability theory were connected with games of chance: gamblers asked mathematicians to estimate their chances of winning or losing. To that end, the mathematicians had to find the rules of handling probabilities, and to develop specific methods for calculating these and related quantities. They succeeded, but the important question of the meaning of probability had remained open. All attempts to answer it seemed to fail for some reason or other. Mathematicians finally gave up such attempts as a hopeless task and tried to construct probability theory purely axiomatically as a specific branch of mathematics. This construction excludes, however, any possibility of establishing a direct relationship between probability theory and practice. This is accepted by all mathematicians, as well as the fact that no real problem requiring probabilistic treatment can be solved within the framework of axiomatic probability theory—a fact which is characterized in the theory itself as the incompleteness of the theory.

Not all scientists, even among the mathematicians, are reconciled to this state of affairs. Some mathematicians who do not specialize in probability theory believe, for example, that probability theory cannot be a purely mathematical discipline, if only because of the fact that the well-known paradox suggested by Bertrand does not find its solution in this theory. This matter is highly germane to our discussion, and will now be considered in brief. Here is the paradox. Suppose one wants to find out the probability that the length of a randomly selected chord of a circle exceeds the side of an equilateral triangle inscribed in the circle (see Figure 9). Bertrand had demonstrated that this problem, paradoxically, could be solved in many different ways, whereas the solution had to be unique. (The settlement of this paradox is given in Section 5.1.)

Because of the aforementioned defects of axiomatic probability theory, many specialists, including most physicists, prefer using various practical definitions of probability, which are not sufficiently rigorous from the mathematical standpoint.

Hence, the situation with the concept of probability is quite paradoxical as well. The experts in probability theory, who master the means for calculating the quantities involved, have no idea of the meaning of the concept. Recall, too, that 'probability' is also a key word in the interpretation of quantum physics, and this obliges us to face the *riddle* of probability. Hence, to get a clue to the correct understanding of what quantum physics is, it is necessary, in the first place, to clarify the enigma of probability.

Our goal now is to try to achieve an explicit definition of probability. To that end, let us examine the connection between probabilistic predictions and practice. The question of probability arises when there is lack of confidence in the 'result' of performing some random test. 'Random' means that no particular result may get preference over any other possible one. Tossing a coin, throwing dice, or raffling a number are examples of random tests. When tossing a coin, one may wonder whether the head or the tail will appear on its upper side. When throwing dice, one may wonder which of the six numbers $1, 2, \ldots, 6$ will appear on their upper sides. And so on.

The question of probability is posed typically in the following form: what is the probability of this? Hence, one always speaks of the probability *of* something. But what does the question refer to? The probability of what is implied? To be able to answer these questions we will consider a simple example.

2.1.1 *Example 1*

Many people have seen, perhaps on TV, a lotto raffle. A possible way of performing it is as follows. Balls marked by different digits are withdrawn at random, one by one, from a specially equipped container, the withdrawn ball being put back after each withdrawal. Suppose there are 20 balls in a definite container, which differ from each other not by a digit mark, but by either red or blue colour and either L or T marked on them. For short, the red balls marked by L will be called RL-balls, the blue balls marked by T will be BT-balls, and so on.

In the container, let there be 8 BT-balls, 4 RT-balls, 3 BL-balls, and 5 RL-balls. One can pose, say, the question: what is the probability that a ball withdrawn at random from the container will be blue? Here the term probability evidently implies the degree of confidence that the indicated accidental result will occur. But the container can yield exclusively what it contains, and only a ball located in it can be withdrawn. Hence, the question can be re-stated as follows: what is the probability that a ball in this container is blue? This question clearly refers to 'a ball in this container', and it asks about the probability that its colour is blue.

But what is 'a ball in this container'? It is definitely not any of the 20 individual balls contained there because each of them has its own definite colour. The question of *probability* of a colour is irrelevant for them, and the only relevant question might be: of what colour is this or that particular (concrete) ball? Then, again, what is meant by 'a ball in this container'?

Some readers may need to make a special effort to understand what now follows. Suppose you (I mean an individual, concrete reader who right now is reading these lines) are standing by the container pending the result of the ball withdrawal. It does not matter whether or not you are aware of the factual composition of the contents of the container, but you know that it contains balls. How do you imagine a ball? If you have never seen any balls, you would find it hard to imagine one. But you did see a ball more than once, and you think of it as something of a spherical shape. This unvizualizable 'something of a spherical shape' stands for any *concrete ball* really existing in the world and can be called 'abstract ball'. Hence 'a ball in this container' represents any particular ball in this container—it is the abstract ball corresponding to the totality (set) of all the 20 individual balls in it.

We have thus arrived at a very important conclusion: that the question of probability stated above refers to the abstract ball corresponding to all the 20 concrete balls in the container. Similarly, the question 'what is the probability that a red ball in this container has the mark T?' refers to 'a red ball in this container', that is, to the abstract ball corresponding to all 9 concrete red balls in this container. And so on.

Generalizing the conclusion in the above example, we may contend that the question of probability always refers to some abstract object corresponding to a particular set of concrete objects (in the present example the concrete objects being balls). The concepts of concrete objects and abstract objects will be considered in more detail in Sections 2–4. However, right now we shall try to find out the meaning of probability. To that end we turn again to the above example.

Even the least mathematical of readers may have guessed the answer to the question 'what is the probability that a ball in this container is blue?', equivalent to the question 'what is the probability that a ball withdrawn at random from the container will be blue?'. Very likely the answer will be: the probability sought equals the ratio of the number of blue balls in the container to the total number of balls there, viz., 11/20. A question now arises: is it possible to verify the correctness of this prediction? But obviously this may be possible only if one knows how to measure probability, that is, how its value can be found in practice. To do that, however, one seems to have to know at least what probability is. The situation becomes thus paradoxically entangled. To cut the Gordian knot we shall try to employ the method of independent random tests. Here is how it works in the above example.

Suppose the balls in the container are thoroughly mixed up and a single ball is withdrawn from it by some special device assuring the randomness of the test. This ball could be any of the 20 concrete balls it contains, each of which has a definite colour and carries a definite mark (T or L). The result of this first random test is recorded by indicating only the colour and mark: no other properties of this ball are of interest in the problem under consideration. Withdrawal of the ball has led to a change in the content of the container. To restore it, the withdrawn ball is put back into the container, and the balls are again thoroughly mixed up. These random tests are repeated as many times as one wishes, and each time the colour and mark of the withdrawn ball are recorded, the ball is put back into the container, and the balls are thoroughly mixed up. These random tests are independent of each other, for the result of any test does not depend on the outcome of the previous ones.

The accumulated records constitute the so-called *statistical data* concerning the colours of and marks on the concrete balls in the container. A question now arises: how can they be linked to probabilities, when we do not know what probability is?

That is a good question. It demonstrates that statistical data obtained in practice cannot be connected with probabilities, unless probability is defined explicitly.

Out of all the old attempts to define probability it is worth mentioning here only the following two which are related to the *frequency* of a random event, which is the ratio of the number of random tests for which the event is the result to the total number of random tests performed.

Many people understand by the probability of an event the frequency of its appearance in a very large number of random tests, viz., the so-called *statistical probability*. Yet this merely experimental definition of probability is not good enough, for its value changes when the total number of random tests changes, to say nothing about its other disadvantages. Von Mises's definition of the probability of an event as the limit of the frequency of its appearance as the number of random tests tends to infinity fails too, for there cannot be such a limit in principle. The general theoretical proof of this statement can be found elsewhere.[4] Here it will be illustrated on the above example of balls in a container.

For every series of random tests, the number of appearances of a blue ball is not predetermined, whereas the total number of tests is a constant for each series. Therefore, generally, for every series of random tests, the frequency of an event is nápredetermined, which prevents it from tending to any limit (including 11/20, the suggested probability of the event).

Thus the problem of how to define probability explicitly remained open. At the same time, the very supposition that there can be a link between statistical data and probability leads one to suspect that experimental statistics, which deals with the accumulation of statistical data, and probability theory, which should be engaged in theoretical probabilistic considerations, may be merely, respectively, the experimental and theoretical parts of the unique science of probability. This guess has proved to be correct, and this science has been called *probabilistics* by me.

Every particular science explores specific phenomena. Probabilistics is no exception. The basic phenomenon of probabilistics, as revealed by the above example, may be described as follows. When a very large number of independent random tests have been carried out on the set of balls in the container, the frequency of the appearance of a ball of a definite colour and mark, that is, the statistical probability of this event, is approximately equal to the relative number of those particular balls. In particular, the statistical probability of the appearance of a BT-ball equals approximately 8/20, of an RT-ball it is ~ 4/20, of a BL-ball ~ 3/20, and of an RL-ball ~ 5/20. These numbers may be taken as the theoretical probabilities of the respective events.

To be able to extend our findings in the above example to the general case, it is convenient to introduce the concept of *classes*. This concept will be explained in more detail in Section 2.3. Here we mean by a class a totality of concrete balls of one and the same kind. We thus have in our example four classes: BT, RT, BL, and RL, corresponding to the BT-balls, RT-balls, BL-balls, and RL-balls, respectively.

It is possible now to give a reasonable and explicit definition of probability for our particular example, namely: 'by the probability that a ball in the container belongs to a definite class is meant the ratio of the number of balls in this class to the total number of balls in the container'.

That is an explicit definition of probability as an objective quantity characterizing the composition of the content of the container. At the same time, this quantity can also be taken as the theoretical probability of the event which is the appearance of a ball belonging to the corresponding class in a random test.

Among other lessons we can learn from our particular example, it is worth emphasizing two. (i) The statistical probabilities of events, that is, the frequencies of the appearance of the events in a very long series of random tests, can be regarded as measured (and, hence, always only approximate) values of the corresponding theoretical probabilities. (ii) Unlike concrete balls, which are the really existing particular ones dealt with in statistical experiments, abstract balls referred to in probabilistic considerations do not exist in reality; thus concrete balls and abstract balls are quite different, and the distinction between them should be kept firmly in mind.

Our task now is to extend the above definition of probability to the general case. We observe again that probability refers to an abstract object, 'a ball in this container' in our particular case, and it concerns the plausibility of the assumption that its colour and mark have some definite values (blue and L, for instance, or something else). In the general case, probability should refer also to an abstract object and concern the plausibility of the assumption that some of its properties have values which determine a definite subset of the whole set of concrete objects, namely, the one in question. The definition of probability in the general case thus requires, first of all, a thorough consideration of the concepts of concrete objects and abstract objects. These considerations, as we shall see, allow us to settle immediately the cat paradox, the chicken–egg paradox, and many others related to quantum mechanics, in particular.

2.2 Concrete objects

We shall start by considering a few more examples. Let me ask you, the reader, 'who is John? (or who is Barbara?)'. 'Your questions are too strange,' may be your reply: 'which John? (which Barbara?)'.

This possible reply is indeed correct. The questions are too vague and cannot be answered, for there are a great many male and female persons with the first names John and Barbara, respectively, who lived in the past, or are living now, and the questions do not indicate which individual persons are meant. It is clear that, in order to be answerable, such a question should refer to some definite person, that is, to a *concrete* person.

The above questions are examples of ones which are answerable in principle when they refer only to definite objects. Other examples of such a situation are given by the questions: 'what colour is a ball?', 'what is the weight of a cat?', 'what is the age of a child?', and so on. These questions, too, are unanswerable in principle, unless they refer to definite objects, that is, to the corresponding *concrete* objects.

We have thus been brought again to the necessity of introducing the concept of concrete objects. Readers may request that I give an exact definition of this concept, but this just cannot be done, since the concept is a primary one, which it is impossible to reduce to anything simpler. Hence, we can only explain what we mean when speaking of concrete objects.

Concrete objects are perceived by us in the most usual, trivial sense, namely, as definite, actually existing objects, which are the subject of any direct examination (observation, measurement, etc.) or of any other kind of direct action on them. Examples of concrete objects are: every reader of this book, each of its copies, every page in that copy, every individual car, and so on.

Some readers may dislike the word 'concrete' for some reason or other. But 'concrete object' is a scientific term and, as such, cannot be replaced by other words. Its meaning as a really existing particular, fully defined object is clear, and anyone who hates the word 'concrete' is free to try to use another word or a combination of words. In this book, I use the term *concrete object* where necessary, expressing it differently where advisable.

In order to specify clearly the concrete object meant, one can use, generally speaking, different methods. In the simplest cases it is sufficient to point out a concrete object in any arbitrary way (even with a finger). In the general case, however, it is necessary to use the properties (characteristic features) of concrete objects. For instance, if there are two chairs in a room, e.g., a low and a high one, it will suffice to say the 'low chair' to specify clearly the concrete object implied. Here we make use of the height property of the chairs which are distinguished by the 'value' of this property, one chair being low and the other high. Thus, we have differentiated between the properties of concrete objects and the values of these properties, and the strict difference between these two concepts should always be kept in mind. If different concrete objects possess one and the same property, they may differ in its values, as in the above example. Every property thus has some number (more generally, some set) of values.

If two chairs are of one height, but of different width, then to indicate the concrete chair meant, it is sufficient to use its width property and say, for instance, 'the wider chair'. The values of the width property here are 'wider' and 'narrower' or 'wide' and 'narrow', essentially the same. Even if two chairs have the same values of all their own properties (which is unrealistic, by the way), they can be differentiated by their position in space; a concrete chair can be singled out by specifying the place it occupies in the room. Here we use the spatial position property of a chair, its values being 'places occupied by a chair'. The 'time' property may also be used to differentiate concrete objects. For instance, if several drops of water are released from a pipette, each concrete drop can be characterized by the corresponding value of a time property: the 'instant' of its fall.

We can summarize what we have learned from the above, and generalize the results by stating that: (i) any concrete object possesses a set of properties; (ii) each property has a set of values; (iii) each concrete object has one and only one value of each of its properties; and (iv) different concrete objects differ in the value of at least one of their properties.

When making practical use of this statement for solving a particular problem, however, it is unnecessary, to consider all the known properties of the concrete objects under consideration. The very statement of the problem should suggest the (least) set of properties to be chosen so that different concrete objects may be discriminated; the objects need to differ, of course, in the value of only one of their properties.

For instance, if one wants to find out how many balls of different colours are contained in some box, it would suffice to distinguish different concrete balls of one and the same colour by the different numbers somehow assigned to them. The temperature, for example, and other specific properties of the balls, which also are likely to differ for different concrete balls, are of no interest in this problem and should be omitted.

Finger-prints are an example of a property that can be sufficient for solving complex practical problems (in criminology). Also, the property 'full name' (first, middle, and last names) of a person may happen to be sufficient in some cases for discerning particular people but insufficient in others. If in a group of people there is only one who has the required value of 'full name', then this particular person can be singled out among all the others (values of the property 'full name' are combinations of the possible values of the properties: 'first name', 'middle name', and 'last name'). However, if the group of concrete people is rather large, it can include a number (a smaller group) of people with the same value of the property 'full name'. In this case, in order to distinguish between different concrete people, it is necessary to have at least one more property, whose value is different for each of them. It is a lack of additional information concerning properties whose values are different for different concrete persons which sometimes leads to misunderstandings. We may recall, for instance, bills being sent to persons who had nothing to do with them, or people arrested by mistake. This sometimes results from the insufficiency of the data provided by computers for the identification of an individual.

A definite value of the property 'full name' therefore determines a group of individuals in some wider group of them, for instance, in the group (in the set, one may say) of all the inhabitants of a particular city.

Similarly, in a set of some concrete objects determined somehow, there are subsets of concrete objects, such that each of them consists of all the concrete objects that have the same values of certain properties. This brings us right to the important concept of classes again (cf. Example 1, in Section 2.1).

2.3 Classes of concrete object

A set of concrete objects having the same values of their properties will be called a 'class' (the set is, as a rule, a subset of a wider set containing more concrete objects). This notion is used in mathematics with the same meaning as here. It also includes the conventional concept of a class used in the social sciences, as a particular case. Each concrete object contained in a set of them may be called an 'element' of the set, and in what follows we shall use the term element frequently. Let us now consider some more illustrative examples.

2.3.1 *Example 2*

A concrete person's wallet contains 18 twenty-dollar, 5 ten-dollar, 12 five-dollar, and 4 one-dollar bills, plus 2 one-dollar coins, 2 quarters, 2 dimes, 3 nickels, and 8 pennies. According to the two values, 'bill' and 'coin', of the property 'kind' of money, the content of the wallet may be divided into two classes: (a) class of bills (39 elements) and (b) class of coins (17 elements). The property 'denomination' (the conventional term value for bills must be ignored in this particular example because its use would introduce a second meaning for value and the ambiguity would be unhelpful.) has the values $20, $10, $5, $1, 25¢, 10¢, ¢5, and 1¢. The division of the money into the corresponding 8 classes gives: (a) class of $20 (18 elements), (b) class of $10 (5 elements), (c) class of $5 (12 elements), (d) class of $1 (6 elements), (e) class of 25¢ (2 elements), (f) class of 10¢ (2 elements), (g) class of 5¢ (3 elements), and (h) class of 1¢ (8 elements).

If we use both properties for the division of the money into classes, we get, first, the corresponding 9 combined values: (bill, $20), (bill, $10), (bill, $5), (bill, $1); (coin, $1), (coin, 25¢), (coin, 10¢)), (coin, 5¢), and (coin, 1¢). The respective classes are: (aa) class of $20 bills (18 elements), (ab) class of $10 bills (5 elements), (ac) class of $5 bills (12 elements), (ad) class of $1 bills (4 elements); (bd) class of $1 coins (2 elements), (be) class of 25¢ coins (2 elements), (bf) class of 10¢ coins (2 elements), (bg) class of 5¢ coins (3 elements), and (bh) class of 1¢ coins (8 elements).

Each piece of money also has some geometrical, physical, and other properties, in accordance with whose values (and their combinations) the money (this set of concrete objects) may also be divided into classes.

2.3.2 *Example 3*

A group of people in a certain room consists of 124 students and one professor. We are dealing, hence, with a set of 125 concrete people. Let 70 students be women and 54 men, and let the professor be a man. Then we have one inherent property of these people already, 'sex', according to whose values, 'female' and 'male', this set can be divided into two classes: (a) class of women (70 elements) and (b) class of men (55 elements).

The property 'age' inherent in people allows us to divide this set into classes by using its values. Let us choose the conventional evaluation of age (for adults) in whole years, and let the numbers of students having the age of 18, 19, 20, and 21 years be 56, 37, 24, and 7, respectively, and professor's age be 28 years. Then the set of people under consideration is divided into 5 classes: (a) class of 18 year olds (56 elements), (b) class of 19 year olds (37 elements), (c) class of 20 year olds (24 elements), (d) class of 21 year olds (7 elements), and (e) class of 28 year olds (1 element).

We may also choose the combined values of both 'sex' and 'age' properties for the classification of the people in this example. Then, in accordance with the possible combined values of these properties, we obtain the following 10 classes, into which they should be divided: (aa) class of 18 year old women, (ab) class of 19 year old women, (ac) class of 20 year old women, (ad) class of 21 year old women, (ae) class of 28 year old women; (ba) class of 18 year old men, (bb) class of 19 year old men, (bc) class of 20 year old men, (bd) class of 21 year old men, and (be) class of 28 year old men. Let us note, by the way, that class (ae) contains no elements, in other words, it is *empty*. Nonetheless, we have to include it in the list of *all* the possible classes. It can happen that some other classes in our example are empty as well. For instance, if there are no men under 20 years of age in this group of people, then classes (ba) and (bb) also are empty.

The values of first, last, and full names may also be used for the division of a set of concrete people into classes. If in this example there are several persons with the same first name, say John or Barbara, then the set of concrete people under consideration contains a class of Johns and/or Barbaras.

There are many other inherent properties in people, such as height, weight, blood pressure, etc., whose values may be used for the division of a given set of concrete people into classes. In addition, there are some specific properties which may also be used for the classification of the concrete people in our example, that is, for their division into classes, such as, in particular, their personal and social ones.

With the foregoing preparation, we should now be ready to turn to a general consideration of the concept of *abstract objects*, whose particular case was discussed in Section 2.1.

2.4 Abstract objects

The concept of abstract objects, in contrast to that of concrete objects, is a secondary one. Its strict definition can be found elsewhere. [1, 4] Here it is presented in a more popular form, and we start with the examples just considered.

In Example 2, class (bh) consists of 8 concrete pennies. They can be discerned from each other somehow. If not, how can one be aware of the number of pennies in the wallet? However, if we say 'a penny in the wallet', we are not referring to some particular, definite coin therein (recall the considerations concerning 'a ball in this container' in Example 1, Section 2.1). We are referring, in fact, to something that stands for any concrete coin of denomination 1¢ in the wallet. Hence, 'a penny in the wallet' represents the whole class (bh) of concrete pennies in the wallet. We may say that 'a penny in the wallet' is the abstract object corresponding to the class (bh) of concrete pennies—a particular instance of concrete objects—contained in the wallet.

On the other hand, 'a penny in the wallet' is the combination of the values of the properties 'kind of money' and 'denomination' which determine the class (bh). Therefore, generalizing, we define the abstract object corresponding to a class of concrete objects as the *totality of the values of properties which then determines the class.*

(In some of my previous publications, 'abstract object' was defined in a seemingly more visual manner, namely as the 'image' of the corresponding class of concrete objects. But this definition is not accurate enough and may cause confusion. Indeed, the image of an abstract object cannot be created in one's mind in principle, due to the fact that the latter does not exist in reality. Only concrete images, which are those of concrete objects, may appear in the imagination of an individual. Nobody can imagine, for instance, an abstract 'professor', or an abstract 'worker', or an abstract 'animal'. At the same time, the wordy description of these abstract objects, actually by the totality of the values of properties which then determines the corresponding class of concrete objects, can be found, in particular, in appropriate dictionaries.)

The name of an abstract object is easily derived from that of the corresponding class of the concrete objects. In this example, 'a penny in the wallet' is the abstract object corresponding to any element of the class (bh) of concrete pennies there. 'A $10 bill in the wallet' is the abstract object corresponding to any element of the class (ab) of concrete $10 bills there. 'A coin in the wallet' is the abstract object corresponding to any element of the class (b) of concrete coins in the wallet. And so on.

In Example 3, 'a woman in the group' is the abstract object corresponding to any element of the class (a) of concrete women there. 'A 21 year old man in the group' is the abstract object corresponding to any element of the class (bd) of 21 year old concrete men there. If there is a class of Johns or one of Barbaras in the group, then 'a John in the group' or 'a Barbara in the group' is the abstract object corresponding to any concrete John or Barbara in this group. And so on.

Note that the name of an abstract object determines it in full. It indicates, in particular, the set of concrete objects whose class of elements the abstract object conforms to. Therefore, as distinct from 'a woman in this group', which is related to the group of concrete women in Example 3, 'a woman in the USA' is the abstract object which conforms to any concrete woman in that country, irrespective of her other individual features and the time she lives. 'An elephant' is the abstract object corresponding to any element of the class of concrete elephants, regardless of where and when they lived, live, or will live. And so on.

Thus, abstract objects related to one set of concrete objects can be of different degrees of abstraction. Indeed, in Example 2, the abstract object 'a bill in the wallet' is more abstract than 'a $1bill in the wallet'. In Example 3, 'an 18 year old man in the group' is less abstract than 'a man in the group'. For the set of all the concrete animals on Earth, the abstract object 'a mammal' is less abstract than 'an animal', but more abstract than 'an elephant'. And so on. This fact demands the careful use of appropriate names when talking about abstract objects.

Now that we have introduced the general concepts of concrete objects and abstract objects, we could continue our movement towards the general definition of probability. However, the immediate corollaries resulting from these concepts allow us right now to settle the chicken–egg paradox, the cat paradox, and many other paradoxes, including some quantum-related ones, even before we have obtained a full awareness of how to define probability and what quantum physics really is. As we shall see, the settlement of all those paradoxes is conditioned by merely keeping firmly in mind the necessity of a strict distinction between concrete objects and abstract objects. Accordingly, we postpone for a while the consideration of general probability and quantum physics, and proceed to a discussion of the necessity to avoid any confusion between concrete and abstract objects.

2.5 The key principle

Some readers may say (I certainly mean concrete readers; an abstract 'reader' who does not exist in reality can say nothing!): 'I have learned almost nothing new or significant from your discussion above. It goes without saying that there exist concrete objects which differ from abstract objects, and even some ancient thinkers were aware of this fact. In some modern works on sociology we can also find mention of the difference between concrete things and abstractions'.

I may agree with such remarks, but were the ideas of concrete objects and abstract objects really clear enough? Was the distinction between them stated sufficiently strictly? That is not the case, as one can see from the fact that confusion about concrete objects and abstract objects is widespread, and it is just this confusion which, not infrequently, is the principal source of misunderstandings and unceasing debates on many issues in both the natural sciences and the humanities, as well as a cause of many misconceptions in everyday life.

Therefore, since it is necessary to comprehend fully the idea of concrete objects and the strict definition of abstract objects, we should dwell on the relationship and distinction between concrete and abstract objects a little longer.

Every concrete object does exist in reality, having definite values of all its properties. An abstract object does not exist in reality—it merely is, and is described by, the combination of the common values of the properties which single out the corresponding class of concrete objects. Everything really existing in the world is concrete, in particular, concrete people deal with concrete things in practice. A concrete bill one receives from a concrete person or a concrete organization can be paid by a concrete cheque (check), concrete money, or in some other appropriate concrete way. Concrete people eat concrete meals, wear concrete clothes, read concrete books, live in concrete places, etc., etc.

Furthermore, abstract objects are meant to have definite values of some properties of the corresponding concrete objects, these values determining their names, but they are not supposed to have definite values of any other properties of those concrete objects. It is these two principal features of abstract objects—the nonexistence in reality and the lack of definite values of many properties—which differentiate them from the corresponding concrete objects. A significant difference between concrete objects and abstract objects, arising from the former, is related to the type of questions that can be asked about them.

Any questions concerning the values of properties of concrete objects are permissible in principle, because the properties of concrete objects do have definite values. Such questions can be relevant for an abstract object only if they refer to those properties whose values determine the name of the abstract object, and are irrelevant for it if they refer to any other properties of the corresponding concrete objects.

In Example 2, we may ask, in principle, any questions about the values of properties of particular bills and coins in the wallet. We may ask, for instance, 'what is the weight of this coin or of this bill in the wallet?' because they do have definite values of weight, even if it is impossible to know them exactly. But we may not ask: 'what is the weight of "a coin in the wallet"?' or 'what is the weight of "a bill in the wallet"?', for these abstract objects do not have definite values of the property 'weight', which therefore renders these questions meaningless.

It might seem that different concrete dimes, for example, should have one and the same weight and, therefore, the same value of 'weight' may be ascribed to the corresponding abstract object—'a dime in the wallet'. In fact, however, that is not the case, since concrete dimes have only *approximately*, not exactly, the same weight, and this is the distinction. The same is true for all the other concrete coins or bills of one denomination. On the other hand, the values of the property 'denomination' are the same for all the quarters, dimes, and pennies, respectively. Therefore, the corresponding definite values of that property are reflected in the names of the related abstract objects.

In Example 3, one may question, in principle, the values of properties of any particular person in the group. One may ask, for instance, 'What is the full name of this student?' or 'of what colour is the hair of that student?', and so on. (Note that for brevity, the conventional way of questioning is being used here, namely, 'what is the full name . . .?', instead of 'what is the value of the full name . . .?', and 'of what colour is . . .?', instead of 'what is the value of the colour . . .?'. Whenever such a simplification should cause no misunderstanding, that is, confusing the properties with their values, I may utilize it in what follows.) Clearly, however, one may not ask 'what is the full name of "a woman in the group"?' or 'of what colour is the hair of "a 21 year old man in the group"?', for the abstract objects 'a woman in the group' and 'a 21 year old man in the group' do not have definite values of either 'full name' or 'hair colour'. At the same time, the questions: 'what is the sex of "a woman in the group"?' or 'what is the age of "a 21 year old man in the group"?' are reasonable because those abstract objects have definite values of the pertinent properties, which determine their names.

It is appropriate at this point to address some possible doubts concerning the above considerations. The statement that abstract objects have these or those values of some properties may seem dubious to a critically thinking reader, who might say: 'you yourself have repeatedly contended that abstract objects do not exist in reality. How can something that does not exist in reality have any values of any properties?'.

The doubt is certainly quite legitimate if we are to take literally a sentence stating that an abstract object has some values of some properties. Such a sentence, however, is rather a figure of speech and means only that the abstract object in question is determined by the indicated values of the relevant properties.

Confusion of concrete and abstract objects does not necessarily lead to grave misunderstandings, but regrettably in many cases it gives rise to unceasing debates, due to the fact that the debaters actually are unaware of what they are talking about. This has happened, for instance, in physics, where many issues related to quantum mechanics in particular have already been debated for many decades. Indeed, debates on some vitally important problems of social life in the humanities have lasted for centuries.

It may be understandable now why a *key principle* (in all my previous works it was called key methodological principle) *must* be introduced and firmly kept in mind. This principle reads: *it is necessary to distinguish strictly between concrete objects and abstract objects.*

It is worth noting that one can arrive directly at this principle in the following simple way. As was mentioned in the preface, any science has both theoretical and experimental parts. The subjects of the experimental part are really existing individual objects, on which experiments are carried out, that is, the ones we call concrete objects. The subjects of the theoretical part are meant to be void of the individual features that distinguish concrete objects from each other, and are inessential and unnecessary in theoretical considerations. They, obviously, are ones we call abstract objects. We may now say that concrete objects are those we actually deal with in practice, whereas abstract objects are ones we use merely in our deliberations and discussions. Since these two kinds of subject, the respective parts of a science, are entirely different, the strict distinction between them is an obligatory requirement.

This principle should be taken into account not only when constructing any science, including a humanistic one, but also in our everyday life. An elementary example of this is given by the answer to our earlier questions of who you are and who I am (see Preface). You are, in truth, the concrete person, a particular instance of a concrete object, who right now is reading these concrete lines. You are not an abstract person without a name, a sex, an age, a nationality, and so on. I am the concrete person who has written these concrete lines, and the concrete author of this concrete book. Like you, I am not an abstract person—'the author'—without a name, a sex, an age, or nationality.

We now apply this principle, in order to demonstrate how it works, to some simple examples of paradoxes.

2.5.1 *Chicken–egg paradox*

This paradox has been stated in Section 1.3. It is created by the seeming impossibility of answering the question: 'which comes first, the chicken or the egg?'. This question, however, is improper, since it refers to an abstract chicken and an abstract egg. An abstract chicken corresponds to all concrete chickens existing both before and now, and also to those concrete chickens which will exist in the future, and an abstract chicken cannot have a definite 'instant' value of the time property, which is necessarily associated with the 'before' and 'after' concepts. The same is true for an abstract egg. If the question refers to concrete chickens and eggs, the answer is unambiguous and very simple: each particular chicken precedes the particular egg it hatched, but succeeds the particular egg from which it was hatched.

Some people confuse the question 'what *comes* first, the chicken or the egg?', related to the chicken–egg paradox, with the question 'what *came* first, the chicken or the egg?' that has nothing to do with the former. The latter actually concerns the origin of chickens and eggs, and it might rightly be asked in pre-evolutionary theories. Evolutionary theory, however, makes such a question irrelevant.

2.5.2 *Cat paradox*

This paradox was set out in Section 1.2 and some attention was given to it in Section 1.4. The paradox is contained in the conventional quantum mechanics answer to the question: 'is Schrödinger's cat alive or dead during the experiment, before the experimenter has established its outcome?' The answer states that during the experiment the cat is neither alive nor dead—it is 'alive-dead' (or 'dead-alive'), even though the experimenter detects only one result: either alive or dead. The reason for such a conclusion is that, since both outcomes are equally probable, the state of the cat during the experiment is altogether uncertain.

Now, what is wrong with this conclusion? Since the reasoning of conventional quantum mechanics employs probability, it must concern an abstract cat (see Section 2.1), whereas any cat experiment, even an imaginery one, should be related to a concrete cat. But every concrete cat during the experiment has only one of the two possible values, 'alive' and 'dead', of the property 'state of being', whereas an abstract cat does not exist in reality at all. Therefore, the question concerning the 'state of being' of an abstract 'Schrödinger's cat', as stated above, is senseless.

We thus see that the cat paradox results actually from the inadmissible attempt to answer an improper question irrelevant for an abstract 'Schrödinger's cat.' Note, by the way, that the chicken–egg paradox has a similar origin—it too, is caused by posing an improper question.

I shall now turn to a consideration of some other, more complicated paradoxes. (The readers of the next few sections will hopefully be acquainted with the required background in *classical* mechanics. If not, they may pass lightly over these somewhat technical matters, but we must anchor our principles in concrete examples.)

Chapter 3
Some paradoxes in modern physics

3.1. Einstein–Podolsky–Rosen's (EPR's) paradox

EPR's arguments[5]—seemingly contradicting conventional quantum mechanics—are related to a specific version of the following mental (thought) experiment. Imagine a pair of particles each of which moves freely along the same straight line: that is, without being acted upon by any force, so that the distance between them remains constant (see Figure 2). The total momentum of the pair, which equals the sum of the momenta of the two particles, is constant too, due to the *law of conservation of momentum* of classical mechanics. The masses of the particles, the distance between them, and the total momentum of the pair are supposed to be known.

If one measures the momentum of *one* particle, according to EPR, then that of the other becomes known too, because the total momentum of the pair is constant. By measuring the position of one particle, one can easily find that of the other, because the distance between them is constant. Thus, the momentum or position of either particle can be predicted with certainty by making a measurement of the chosen property on its counterpart.

Furthermore (and this was not noticed by EPR), since the total momentum of the pair is known, there is no need to measure the momenta of the separate particles—they can be calculated easily, since both particles move freely with the same velocity (due to the distance between them remaining constant) as the pair (its centre of mass). Hence, EPR's arguments actually lead to the conclusion that the values of the momenta and positions of the two particles can be found simultaneously.

Is anything wrong with this reasoning? I do not think anyone might find any flaw in it, and indeed nobody did until now. But EPR's conclusion seemingly contradicts conventional quantum mechanics which asserts that the pair of particles is described by a wavefunction that gives probabilistic predictions only, and cannot yield exact values of the position or momentum of either particle. It also seems to contradict the conventional quantum mechanics belief that the momentum and position of a particle cannot have

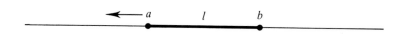

Figure 2. *Two particles a and b are moving freely in the direction indicated by the arrow, the distance l between them being constant.*

definite values simultaneously (widely known as the Heisenberg uncertainty principle).

I continually stress that these are only *seeming* contradictions, because there is no contradiction at all—only a confusion of concrete and abstract particles. EPR's arguments concern a mentally performed *experiment*, and they certainly meant *concrete* particles, as they had to do, because any experiment, even one performed in the mind, can be carried out only on concrete objects. Quantum mechanics, however, involves probabilities which refer to *abstract* particles in the case under consideration. And this alone easily removes the seeming contradiction between EPR's arguments and conventional quantum mechanics. (The meaning and validity of the Heisenberg uncertainty principle will be discussed in Section 6.4.)

It is interesting to note, by the way, that for the particular Gedanken-experiment suggested by EPR, the total momentum of a concrete pair is zero, which means that the pair, taken as a single entity, is at rest. This, together with the requirement for the distance between the concrete particles in the pair to be constant, signifies that both particles are also at rest.

There is one more aspect that we can discuss usefully now. The most paradoxical inference revealed by EPR is as follows. According to conventional quantum theory, the state of either particle in the pair depends on what physical quantity has been measured on its counterpart, no matter how distant the two particles are from each other. This should mean that a particle, even when extremely remote from the other of the pair, immediately 'recognizes' that a measurement has been made on the other particle and what the result was, even though there can be no interaction and no real signal can propagate with a speed exceeding the ultimate rate c of special relativity. This inference has provoked various speculations, including that of 'action at a distance', and the so-called 'spooky' effect, etc.

This paradox, too, is easily removed by recalling that quantum theory has nothing to do with measurements in principle, for it is related to abstract physical systems which do not exist in reality, whereas a measurement being an experimental procedure, can be made on only concrete systems. Regarding the puzzle of how a very distant individual particle can immediately 'recognize' the result of a measurement (say, of the momentum) on the other particle in the same pair: in fact nothing of this kind occurs. A measurement of a quantity on a concrete particle reveals just the value that quantity really has. (The measurements are considered here to be ideal, that is, yielding the exact values of the properties measured.) Now, since the sum of the momenta of the two concrete particles in any one pair is constant, the values of the two momenta are strictly interconnected. Given the momentum of one particle, that of the other (in the same pair) is determined fully by the law of conservation of momentum, no matter how distant from each other the particles are, and any measurement will reveal this fact.

EPR's thought experiment cannot be realized in practice, and its controversial inferences are, hence, unverifiable. Therefore, other experiments have been suggested that in principle are realizable: in particular, some related to Bell's famous inequalities.[14] Most of them involve either the spin of particle or the polarization of a photon. These experiments, undertaken (both in thought and in practice) to verify EPR's conclusions, have naturally been regarded as particular versions of EPR's experiment, but actually they are not, for the following reasons.

There are two types of science: science that needs no probabilistic treatment (probability unrelated science), and probability related science. It is essential that the link between the experimental part of a science and its theoretical part differs in these two types. In a probability unrelated science, the theoretical laws applying to an abstract object are valid, at the same time, for each and every corresponding concrete object, so that the theory can be verified on any one of them. This is the case, in particular, with *classical mechanics*, because of which the mechanical behaviour of every concrete mechanical system can be predicted (more accurately, pre-calculated) with certainty.

In a probability related science, however, theoretical laws applying to abstract objects are of an essentially probabilistic character. They cannot in principle be verified on a separate concrete object—their verification requires the gathering of proper statistical data by carrying out an appropriate statistical experiment on a large enough number of concrete objects. Therefore, in probability related sciences, definite values of properties possessed by a particular concrete object cannot be predicted in principle by theory. This is the case with any probability related domain of physics, in particular with quantum physics.

Now, the concrete mechanical systems of EPR belong to classical mechanics, a probability unrelated (i.e. *deterministic*) science, whereas the concrete systems dealt with in the experiments involving either particle spin or photon polarization belong to probability related sciences, as will become obvious in the following pages. The seeming paradoxes revealed in Bell's inequalities-related experiments are actually connected with the wrong identification of these experiments as particular versions of EPR's experiment. The quantitative discussion of their results involves awareness of some probability related matters. We should, therefore, delay it until the appropriate information has been given (see Section 6.5). However, a qualitative understanding of the results can be gained simply by applying the key principle (see Section 2.5), that is, the principle of strict distinction between concrete and abstract objects. As follows from the considerations that led to the introduction and a detailed explanation of the principle, its application requires a clear understanding of the properties of the concrete objects involved. Hence, before considering spin-related experiments, we shall discuss the peculiarities of the spin of concrete particles.

3.2 Angular momentum and spin

The peculiarities of spin are in a way similar to those of the angular momentum of concrete mechanical macrosystems. We start, therefore, with a discussion of the latter.

It is interesting to compare the angular momentum, which is connected with the rotational motion of a concrete mechanical system, with the momentum of the system, bound up with its translational motion. A finite rectilinear displacement of a concrete rigid body (translational and rotational motions of a concrete mechanical system can be examined on rigid bodies) can always be represented as a (vector) sum of three rectilinear displacements in three mutually perpendicular directions (see Figure 3). Irrespective of the order in which all three displacements are carried out, the resulting displacement is the original one. This fact characterizes the rectlinear displacement of a rigid body as a vector. The three displacements in any three mutually perpendicular directions, which represent the vector, are called its orthogonal components.

Rotational motion is different. A finite angular displacement of a rigid body about any axis cannot be presented as a sum of its angular displacements about any other axes, as can be seen, for instance, from Figures 4 and 5. A stick originally disposed along the x axis becomes finally orientated along the z axis if it first turns through 90° about the y axis and then through 90° about the z axis (Figure 4). If the sequence of the turns is reversed, the final position of the stick is along the y axis (Figure 5). Hence, a finite angular displacement is not a vector.

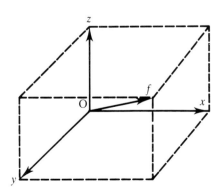

Figure 3. *Three displacements, Ox, Oy, and Oz, in three mutually perpendicular directions result in one and the same displacement Of, irrespective of the order in which they are performed.*

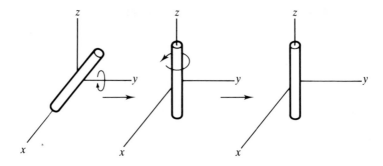

Figure 4. *A stick originally disposed along the x axis becomes finally orientated along the z axis if first turned through 90° about the y axis and then through 90° about the z axis.*

A similar distinction turns out to exist between the corresponding related quantities the momentum and the angular momentum of a concrete mechanical system. The momentum is a vector, whereas the angular momentum is not, because there can be only one (rotational) axis at a time for every concrete mechanical system, and only the angular momentum related to it actually exists at that time.

It should be noted, when speaking of an axis, that its direction is implied. Since the positive and negative directions of a rotational axis are opposite, the angular momenta associated with them are of the same magnitude, but of opposite signs.

In classical mechanics, the angular momentum is defined usually as a

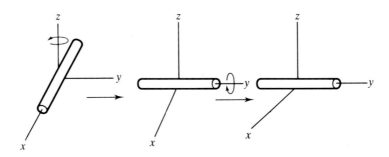

Figure 5. *A stick originally disposed along the x axis becomes finally orientated along the y axis if first turned through 90° about the z axis and then through 90° about the y axis.*

vector referred to some spatial point chosen, in particular, with reference to the centre of mass of the relevant system, and the angular momenta related to some rectangular axes are supposed to be its respective components. However, because of the above mentioned fact that only one rotational axis can exist at a time for every concrete mechanical system, angular momenta related to different axes cannot exist simultaneously, which means that the angular momentum does not exist as a vector either. Incidentally, the false impression of the existence of the angular momentum vector is created by binding it up with a rotation about a spatial point. But such a rotation does not exist in reality. Any so-called 'rotation about a point' is, in fact, one about one particular axis that exists at the particular time instant the rotation takes place.

Angular momenta related to different axes have different values. There always exists, however, at least one axis, such that the angular momentum related to it has the maximum absolute value that coincides with the magnitude of the alleged angular momentum vector.

For a closed mechanical system, that is, for one which is not acted upon by any external force, the conservation of the angular momentum related to any axis chosen can be proved easily. Conservation of its maximum absolute value follows immediately from this. Both conservations laws are fulfilled in common.

The peculiarities of the angular momentum of a concrete closed mechanical macrosystem can thus be set out as follows: 1. a definite value of the angular momentum exists for every axis chosen; 2. there are no simultaneous values of angular momenta related to different axes; 3. there is a maximum absolute value of the angular momentum which refers to at least one particular axis; 4. the conservation law for the angular momentum related to any axis chosen and that for its maximum absolute value are fulfilled in common.

The above peculiarities prove to be similar to those of the spin of a concrete microsystem, in particular of a concrete particle. Spin is the inherent property of every concrete particle, and has the dimension of the angular momentum. All attempts to link the spin of a particle with the rotation of the particle have failed, and I do not think they can ever be successful. In any case, we do not know the origin of the spin, but this does not matter. What matters now is simply that we know its peculiarities. For the particular case of the concrete particles we shall deal with (spin-$\frac{1}{2}$ particles), these peculiarities can be described as follows:

1. The absolute value of the spin related to any axis is 1 (in certain units); 2. only one value of spin (1 or −1) refers to every axis chosen; 3. there are no simultaneous values of spin for different axes; 4. the total spin of a closed system of concrete particles, related to any axis, is constant, viz., it does not change in time (the *spin conservation law*).

For the same reason as in the angular momentum case, the spin values referring to the opposite directions of an axis are of opposite signs.

Keeping in mind the above peculiarities of spin, we can proceed to a discussion of some controversial experiments, both imaginary and real, whose results seem paradoxical in one way or another.

3.3 Bohm's paradox

The gist of Bohm's arguments[6] can be expressed as follows. The total spin along any axis of a system consisting of two spatially separated spin-$\frac{1}{2}$ particles is zero. This means that if one particle has spin 1 along some axis, the other has spin −1 along the same axis, or vice versa. One cannot predict the result of a measurement of the spin of either particle along, say, the z axis. However, if this result turns out to be, for instance, 1 for one particle, then the result of a measurement of the spin of its counterpart along the same axis can be predicted to be −1. Similarly, the result of measuring the spin of either particle along *any* direction can be predicted by the experimenter, with certainty, after the result of measuring the spin of the other particle along the same direction has become known. Thus, the values of the spin of each particle along any direction are definite, whereas quantum mechanics contends that this is impossible.

This case is more complicated than that set out by EPR, for the latter deals with the well known peculiarities of momenta and coordinates, whereas the former involves spin, with its less well known specific features. Therefore, we should be very careful when examining Bohm's arguments. First of all, there is again the confusion of concrete and abstract objects. Any discussion of measurement implies that the objects involved are concrete, whereas any mention of quantum mechanical concepts implies that only abstract objects are under consideration. Hence, in Bohm's case, the possibility of predicting (more exactly, calculating) the result of the measurement of the spin of one *concrete* particle along some direction if the spin of its counterpart along the same direction is known, is determined just by the law of spin conservation. But is it possible to verify the 'prediction' by an appropriate measurement? This is a good question because of the specificity of the measurements intended to verify predictions.

One should be cautious when attempting this, for both measurements are to be made on one and the same pair of concrete particles. If the measurements are gradually made on each of many pairs of concrete particles which have turned up at random, as is the case when verifying probabilistic predictions (see Section 2.1), it is necessary to avoid confusing particles belonging to different pairs. To that end, both measurements must be made in *one* random test, that is, practically simultaneously.

For the same reason, measurements could not in any way yield different 'spin components' of one concrete particle, that is, its spin values along different directions, even if they existed. But the point is that they do not, as follows from peculiarities 2 and 3 of spin. Simultaneous different components of the spin of a concrete particle, I must stress once again, do not exist in reality: a concrete particle has only one spin component at a time, which is related to the particular axis chosen that time.

The real experiments related to the Gedanken-experiment suggested by Bohm were carried out in connection with the famous Bell's theorem, which will be touched upon in Section 6.5. Their results have confirmed that the spins of two spatially-very-remote-from-each-other concrete spin-$\frac{1}{2}$ particle belonging to one pair, whose total spin was initially zero, have opposite values, regardless of the direction along which they have been measured. These results have again provoked the same puzzling question, as in EPR's case: how could one particle know of the result of the spin measurement for the other one? Again, speculations involving 'action at a distance', 'spooky', 'many worlds', etc., have been offered to answer that question.

But those experimental facts have nothing to do with the 'particle's knowledge' of the measurement results. Again, it is just a conservation law, this time the spin conservation law (see peculiarity 4 of spin), which determines that the spins of the two *concrete* particles under consideration *must* have opposite values along any direction. The measurements, which are supposed to yield the exact values of the measured properties, merely reveal that fact.

Similar real experiments on photons by Aspect (and his teams)[7] involving the polarization of photons, instead of the spin of particles, gave analogous results. Before considering those experiments, a few words on polarization and its peculiarities are in order.

3.4 Polarization of photons

The phenomenon of the polarization of light has been known since the 17th century. The electromagnetic theory of light explains it straightforwardly. A beam of light can be regarded as consisting of electromagnetic waves. The electric and magnetic vectors in every wave are vibrating perpendicularly to both each other and the beam. Natural light consists of waves which have all the possible directions of the electric vector in the plane perpendicular to the beam. When natural light falls on a polarizer, only those waves pass through it whose electric vectors are parallel to a specific straight line in the polarizer, which I will call the *p*-line. The light after the polarizer is said to be *linearly polarized*, its line of polarization (briefly, *polarization*) being parallel to the *p*-line of the polarizer. When one more polarizer is put behind the first one (see Figure 6), the polarized light passes fully through it only when its polarization is parallel to the *p*-line of the second polarizer (the second polarizer is called, for obvious reasons, the *analyser*). If the *p*-line of the analyser is perpendicular to the polarization of the light, no light whatsoever appears behind the analyser.

When the polarization of the light makes an angle α with the p-line of the analyser, the intensity of the light passing through the analyser is proportional to $\cos^2 \alpha$. This law was discovered by Malus in 1810 and acquired his name (the parallel and perpendicular cases obviously are contained in this law). The theoretical explanation of this law is very simple from the standpoint of the electromagnetic theory of light. Only that component of the electric vector which is parallel to the p-line of the analyser passes through it. The amplitude of the component is proportional to $\cos \alpha$, and the intensity of light is proportional to the square of the amplitude. We will have to recall the Malus law when we discuss the quantitative results of Aspect's experiments. In the next section, however, we will confine ourselves to a consideration of only their qualitative results. Since the experiments were carried out on photons, we should first find out what polarization is from the standpoint of the idea that light consists of photons.

Polarization should be regarded now as a specific inherent property of concrete photons—it is just every concrete photon which does or does not pass through a polarizer depending on its polarization. The questions of why and how it happens are not of immediate interest to us when examining the related experiments. It suffices to suggest that the property 'polarization' of concrete photons has two values, which I will call p-values, according to which of two possible behaviour patterns they show with respect to a polarizer. The peculiarities of the polarization of concrete photons can then be expressed in a way analogous to that for the spin of concrete spin-$\frac{1}{2}$ particles, as follows. 1. Given a p-line perpendicular to the direction of a concrete photon motion, the polarization of the photon can have only one of the two possible p-values—parallel to the p-line ($p = 1$) or perpendicular to it ($p = -1$). These numbers (1 and -1) are chosen to conform to Aspect's quantitative results, and also they are the same as in

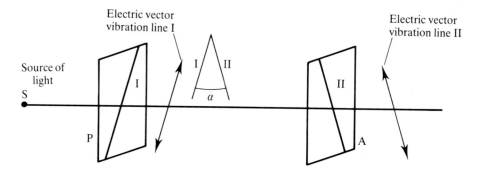

Figure 6. *The electric vector of light that has passed the polarizer P is parallel to its p-line I. Only the component of the electric vector which is parallel to p-line II of the analyser A passes the latter. The intensity of the light after passing through analyser A is proportional to* $\cos^2 \alpha$, *where* α *is the angle between p-lines I and II (Malus law).*

the discussed previously case of spin-$\frac{1}{2}$ particles. 2. Only one p-value is associated with every p-line chosen. 3. There are no simultaneous p-values for different p-lines.

Unlike the spin of a concrete particle, whose values refer to a directed axis, the polarization of a concrete photon involves two mutually perpendicular straight lines in a plane perpendicular to the direction of its motion. If one of them is a p-line, then the other can be named the *complementary* p-line (p_c-line). For a concrete photon, the p-values for these two mutually complementary p-lines have opposite signs.

Keeping in mind the peculiarities of the polarization of a concrete photon, we can easily understand, at least qualitatively, the results of Aspect's experiments, and make sure that they do not lead to any paradoxical conclusions.

3.5 Aspect's experiment

This experiment[7] deals with the measurement of the polarization of pairs of photons, whose constituents are emitted in opposite directions. The measurements are made with the aid of appropriate analysers placed sufficiently far apart from each other. The mutual polarization orientation of the analysers, that is, the angle between their p-lines, is variable (see Figure 7).

The measurements have provided statistical data for the polarization values of photons on the both ends of the setup, related to different spatial orientations of each analyser. The data corresponding to the two photons belonging to one pair have been analysed, in order to find out whether there is a specific correlation between their p-values related to various orientations (for each of the photons) of the relevant p-lines.

It has turned out that the measured p-value of either photon, related to some p-line chosen, is apparently determined by the result of measuring the p-value of its counterpart, related to a different p-line chosen for the latter. It may seem as if the result of the measurement on one photon is conveyed immediately to its counterpart, so that the latter can adjust its p-value accordingly, no matter how distant the two photons are from each other. Is this not paradoxically strange?

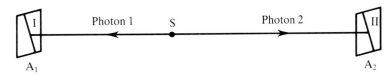

Figure 7. *Two photons of one pair are emitted from one source S in opposite directions. Their linear polarizations are measured by analysers A_1 and A_2, respectively. The angle between p-lines I and II of the analysers is variable.*

To settle this paradox, the same ideas as in the spin-$\frac{1}{2}$ particles case have been tried, such as 'action at a distance', 'many worlds', etc. But this paradox is due, again, in the first place, to the confusion of concrete and abstract photons. The measurements are carried out on pairs of *concrete* photons and reveal the p-values the photons *really* have at the time the measurement has been made. Therefore, the data obtained refer to the real p-values of the concrete photons of each concrete pair examined. Now, the primary result of Aspect's experiment can be put as follows.

If concrete photons of one pair, emitted in opposite directions, have equal p-values for an arbitrary p-line at one time instant, then the same holds true at any time instant as long as neither concrete photon undergoes some action that changes its p-value (that is, as long as the pair remains a closed system).

It is just this 'law of conservation of polarization', similar to the 'law of conservation of spin' for every pair of concrete spin-$\frac{1}{2}$ particles with total spin zero, which is responsible for the so-called 'correlation' between the p-values of the two mutually remote concrete photons belonging to one pair, regardless of the distance between them. And this explanation removes the above paradox in a fully realistic way.

Chapter 4
Personality versus society

4.1 A short pause for reflection

It might be difficult for some readers, I suspect, to digest the material I have already unloaded before them almost without a break. Hence, a short pause may be helpful, which can be used for a brief summary and some reflections.

Our contemplation of Schrödinger's cat led us to the necessity of seeking the meaning of probability. In doing this, we have arrived at the conclusion that it is necessary to introduce the concepts of concrete objects and abstract objects, and to distinguish strictly between them. We have learned that all experiments can be and are carried out only on concrete objects, whereas abstract objects are a subject of theoretical considerations. Therefore, any probabilistic considerations are related to abstract objects, whereas experimental statistics deal with concrete objects exclusively.

We were able to give an explicit particular definition of probability for some special cases considered, and prepare the way for a general explicit definition of probability, which would lead us to an understanding of the essence of quantum physics. But the temptation to use the key principle immediately for settling some paradoxes was so strong that I interrupted our movement towards that original goal and turned to the treatment of these paradoxes.

We have easily settled the 'chicken–egg' and 'Schrödinger's cat' paradoxes, which have nothing to do with any physical considerations, and then went on with more complicated ones involving some specific laws of physics—conservation laws, in particular—which have also been resolved in quite a simple way.

The examples already considered have revealed, once again, the importance of the strict distinction between concrete and abstract objects, while the reasoning that has led to the introduction of and differentiation between these two types of object seems simple and clear enough. Can there be any difficulties with the understanding and recognition of these simple truths? It seems not but, as a matter of fact, there are some.

The idea in itself that there are concrete objects and abstract objects seems rather trivial—a commonplace. This fact alone creates some problems already. Nobody would pay attention to the idea because of its seeming insignificance. Typical was the response to it by Professor B.V. Gnedenko (Moscow State University), a world renowned expert in probability theory and statistics who was also interested in the philosophy of science. When I expounded to him, during several evenings in 1971, the contents of my paper (published later in *Foundations of Physics*, 1973), in which the foundations of probabilistics and probabilistic physics (including quantum mechanics) had been set forth, proceeding from the above ideas, he said to me something like this: 'You know, I had always been aware of that, but I have never thought that it is so important'.

I believe that others who may manifest a slighting attitude towards the necessity for distinguishing between concrete and abstract objects are prompted by all-too-human psychological features. Concrete people, with all the richness of their individual peculiarities, have some common features, such as inquisitiveness and creativity, on the one hand, and an inertia of thinking, and a desire to believe (sometimes blindly) in the correctness of the already established (old) ideas, on the other hand. Only if a special urgent need arises, do they start to consider new, alien ideas.

Among the readers of the present book, there are, I believe, more than a few who are really interested in puzzling out the seeming mysteries related to quantum physics. It has turned out, however, that the necessity for a strict distinction between concrete and abstract objects is a principle that, in the last analysis, not only allows one to establish a proper order in probability related sciences, including quantum physics, classical statistical physics, etc., but also provides a safe basis for a scientific treatment of problems usually considered to be within the realms of the humanities. I trust that some readers will be interested in this side of the story as well.

Here is one corollary of the above key principle, as related to philosophy. Consider the long-standing debates among some philosophers concerning the relationship between matter and spirit (or, in some other versions, between being and consciousness). The materialists contend that matter is primary, whereas spirit is secondary. The idealists' belief is quite the opposite.

Now, what is meant by matter? One of the most renowned representatives of and violent adherents to materialistic philosophy, V. I. Lenin, had defined it as 'the objective reality given to us in our sensations'. This definition suffers, however, from obvious faults. First of all, our sensations deal with concrete things only which, by the way, are not necessarily of a merely physical nature (among them can be particular works of art, for instance, particular musical compositions, performed by particular musicians, particular dreams, particular birds' songs, etc). They all objectively exist in reality, reaching us via our senses, but no reasonable person, I believe, would agree to calling any of these concrete things 'matter'. Even more importantly, no concrete thing our senses can deal with, even of a physical nature, may be called 'matter', in principle: one would hardly say that the concrete table one looks at is 'matter', or the concrete house one owns is 'matter', and so on.

Each and every concrete table, or house, or whatever else, is made of some particular substance (of some particular matter, one may say, utilizing this specific meaning of the word matter). This does not mean, however, that they themselves can be called 'matter'. Matter, hence, should really mean something abstract and, if this concept is to be used, the best way of introducing it is as follows.

Recall the introduction of concrete objects (Section 2.2). It implies that concrete objects possess, among all their properties, the physical ones, in particular mass and, hence, energy. Now, 'matter' can be defined as the most abstract object corresponding to all the possible concrete physical objects in the world. Being an abstract object, 'matter' does not exist in reality. Hence, it cannot be sensed and, second, does not constitute a primary real entity.

Thus Lenin's definition of 'matter' makes no sense, and the basic philosophical premise of Marxism–Leninism (the primacy of matter) is senseless as well. However, this does not mean that the basic premises of any idealistic philosophy are right.

What, then, is right? What can be opposed to both materialism, in particular Marxist–Leninist materialistic philosophy, and any kind of idealism? It is, I believe, a special kind of realism, which I am trying to convey in the present book. The basic premise of this philosophy is the assertion that only concrete things exist in reality. They include not only the (physical) concrete objects previously discussed, but also any related *concrete things* of any nature, such as, for instance, concrete events, interactions between concrete objects, concrete properties, concrete works of art and music, concrete dreams, activities of concrete people, and so on, which are all real as well. And, again, to every set of concrete things of one kind there corresponds a pertinent *abstract thing* that does *not* exist as such in reality and is merely the subject of consideration, discussion, etc. The necessity for strict distinction between *concrete things* and *abstract things* is the (extended) key principle which is a crucial point of this realism.

One may speculate that, had that principle been widely known and recognized many centuries ago, the course of the history of mankind might have been quite different. There might not have been, in particular, any reasonable idealogical ground for many utopias and, especially, the pseudo-scientific, extremely aggressive, dangerous, inhumane utopia of Marx–Lenin, whose virus has infected the minds of many people all over the world, plunged the population of many countries into servitude, destroyed the economics of those countries, and led to innumerable sufferings and the untimely deaths of human beings of several generations. We are eye-witnesses to the general process resulting in the collapse of the pernicious Marxist–Leninist ideology, and of the countries themselves, whose rulers had professed that ideology. Certainly, however, it would have been much better if that ideology had never appeared.

Why has it happened? Why does Marxism–Leninism fail the practicability test?

To answer this question we should realize that Marxist–Leninist ideology is the *theoretical basis* of the communist regimes implemented in many countries, starting with Russia (in 1917), and introduced either by means of deception or by force, or by both. What was going on in the world since the so-called 'Great October Revolution' was actually a vast experiment carried out on many hundreds of millions of living human beings, whose ultimate results were supposed to give happiness and freedom to all the people of the remote future generations. Examination of all the aspects of this subject would require the efforts of many researchers in many fields of the humanities, and is out of place in the present book. But, it is immediately clear that the failure of the Marxist–Leninist historical experiment indicates that the theory it was based on is wrong.

It may sound strange, but the experimental failure of Marxism–Leninism has basically the same origin as the experimental failure of Bell's inequalities (though I certainly have no intention of comparing the immeasurable damage caused by the former with the quite innocent disadvantages of the latter). Both failures result, in the final analysis, from not taking into account the *real* peculiarities of the *concrete* objects on which the experiments have been carried out.

Bell's inequalities involve some probability related quantities, and we shall touch upon them in Section 6.5. However, it is possible to say already that their experimentally proven fallibility is due to the fact that in deriving them the *real* properties of the spin of *concrete* spin-$\frac{1}{2}$ particles (see Section 3.2) and of the polarization of *concrete* photons (see Section 3.4) were neglected.

So what is wrong with Marxism–Leninism? One of its basic fallacies was the neglect of the *real* peculiarities of *concrete* people and dealing, instead, with some far-fetched *abstract* people who do not exist in reality and have nothing to do with real human beings.

In connection with the aforesaid, I would like to emphasize that, even though the principle of a strict distinction between concrete and abstract objects is of extremely great significance for each and every branch of natural science, it is of no less importance for the humanistic sciences, since among the problems supposed to be treated there are vital ones crucial for the very existence of mankind.

In reflecting on the consequences entailed by our consideration of Schrödinger's cat, we have thus come across problems that are (or should be) of utmost interest to all thinking people. I believe it is not out of place, therefore, to touch upon them right now, and I hope that my readers will forgive me this liberty. Those who are not particularly interested in these matters can lightly skip the next few sections and continue with Section 5.1, where I shall go on with the consideration of probabilities and quantum physics.

4.2 Properties of concrete human beings

In treating any social (people-related) problems, one should try, first of all, to achieve a clear understanding of the properties of *concrete* human beings, which are of extreme variety and diversity. In addition to the common properties of human beings regarded as physical objects there are also biological ones, as well as specific human properties of different kinds, which may not even be defined distinctly enough. The values of these last mentioned properties are not unconditional, not quantitative, and in principle cannot be objectively measured with the aid of any devices. (I should remind readers not to confuse 'properties' with their 'values'.) For instance, the values of the properties 'intelligence' and 'goodness' may be introduced and estimated only conditionally; they depend on both the conditional scale chosen and the particular person giving the estimate. We will notice, however, that no matter how the conditional scales are chosen and who evaluates the properties, there arise the corresponding distributions of individuals over the estimated values of the properties (that is, the sets of relative numbers of individuals having those values) for every set of concrete people. It should be remembered, too, that an abstract person, which is the abstract object corresponding to the set of concrete people under consideration, does not have any definite, even conditional, values of properties that do not appear in its name. For instance, the 'New York citizen' cannot be said to be clever or stupid, good or bad, etc., since it does not exist in reality as such and corresponds to the whole set of the citizens of the City of New York, who have more or less definite, even though conditional, distributions over these values of the properties under consideration.

We shall, in particular, be interested in the following important properties of an individual: 1. the ability to think ('intellectuality'); 2. the ability to work consciously; 3. the possession of instincts, wishes and aspirations, inquisitiveness and creativity, senses and emotions; 4. the possession of 'moral qualities', such as 'goodness'. 'honesty', 'responsibility', 'truthfulness', etc, associated with the interrelations among individuals. These human properties, with all the variety of their values possessed by different concrete people, constitute the psychological peculiarities of every concrete person—the person's individuality.

At the same time, every individual almost invariably exists in contact with other people. The most widespread form of such contact is materialized through living together with other people in a society, whatever its kind, starting from a small group of them and ending with a particular State. But in order to be able to live together, people should have the ability to communicate with one another. A language is one of the appropriate means of such communication, one of its principal purposes being the attainment of mutual understanding between different individuals.

The fact that every individual has his own peculiarities of mind often leads to certain problems. Many problems also arise in connection with the contradiction between the individuality of every concrete person and the necessity for the individuals to live together in some society. Before considering some of the problems, we have to stress, however, the following point.

'Man' is the abstract object corresponding to the whole set of all the concrete human beings who have lived, are living now, and will live in the future, regardless of their race, sex, age, residence, and other properties. There are specific properties which allow one to consider some concrete biological entity a man, perhaps even conditionally. Since every concrete man has these properties, among others, so does the abstract 'man'. But 'man', being an abstract object, does not have any definite values of any other properties, including those aforementioned.

To every subset of concrete human beings, which contains individuals with definite values of certain properties, there corresponds an abstract object having the same values of the properties, which may be named by these values. For instance, the human 'female' is the abstract object corresponding to the subset of concrete human beings whose value of the property sex is female, etc.

It is necessary to remember, too, that man, human female, human male, and any other abstract objects corresponding to certain subsets of concrete human beings do not exist in reality as such. Hence, everything which was, is, and will be going on due to the existence of mankind, had, has, and will have bearing on concrete people only, not on the corresponding abstract 'man'.

4.3 Mutual understanding

One of the important problems concerning the activities of concrete people is that of mutual understanding. Libraries may be written on this topic, but some brief remarks are appropriate.

Contemporary languages allow one in principle to express one's thoughts and ideas in words. Different individuals speaking the same language are supposed to understand one another, and that may be the case provided that the matters under discussion are well defined and familiar to all the participants in the discussion. But, even in this simplest case misunderstandings may arise for some reason or other. They may be caused, in particular, by the polysemantic meaning of some words, by like sounding words of different meanings, and by the different interpretation of a correctly constructed sentence by different concrete persons (ignoring jargon and slang).

In more complicated cases, the most widespread cause of lack of mutual understanding—provided that the concrete people involved want to achieve such understanding—seems to be the limitation of the mind of any individual. Every individual possesses definite peculiarities of thinking not only innately but also from the circumstances of their experience. It is often difficult for an individual to express particular thoughts adequately in terms of the conventional common language. Besides, some concepts may be conceived differently by different individuals. This fact may lead to extended, if not to never-ending, discussions, since each of the debaters may try to read his own thoughts into other debater's words. Certainly, a mutual understanding can finally be achieved in principle if the debaters are indeed interested in arriving at the truth, and have enough time to do so.

Unfortunately, however, this is a rare event. There are many causes of mutual misunderstanding between individuals. Among them, the afore-mentioned inner limitation of mind of every individual plays perhaps one of the most important roles.

This feature of the mind of a concrete person, together with the concrete conditions of one's growing up, give rise, as a rule, to certain steady ideas and beliefs which often result in ingrained prejudices and stagnation. These prevent the person even from attempting to comprehend the ideas of another individual. In many cases this may be harmless; in others, for example, when science is involved, such a situation may lead to the inhibition of development.

The above is only a small part of the circle of problems related to mutual understanding between individuals. It is necessary to stress in this connec-tion, however, that the concept of understanding and, hence, also of mutual understanding may concern only concrete people, since 'understanding' is closely associated with 'ability to think' or 'intellectuality', which is the property of every individual. The problem of mutual understanding cannot be raised for groups (sets) of concrete people of any size, including countries, since no group of people can have that property—'ability to think'. The common usage of this term in connection with international relations can only be misleading. It has nothing to do with the aforestated real meaning of the term. One may argue that every individual in a particular country has the property 'ability to think', and does this not mean that the country itself also has the property? The answer is: 'no, it does not'. This property might be ascribed to a particular country (or to any other set of concrete people) only if all of its members would have one and the same value of the property, that is, if they all had the same ideas on everything without any exception. But this is never the case.

We have approached a huge number of interconnected problems related to other peculiarities of concrete human beings regarded as specifically social beings, which will be touched upon briefly.

4.4 An individual and society

Among the specific social peculiarities of an individual are innate proper-ties, such as striving for survival, as well as care for other people, possibly based on the instinct for the preservation of the species. There are also properties which are partly innate and partly acquired due to specific conditions during the development of an individual, such as a person's moral qualities, for instance. It is these properties which to a considerable extent determine one's conduct, that is, one's behaviour, relative to oneself, to other individuals, to the environment, and to the rules and laws of the society in which one lives.

Every individual is a social being by his very nature, because any concrete person must live in order to survive among (or in contact with) other people, that is, in some society, as was mentioned before. But, every individual is unique and has, therefore, his own needs, aims, and ideas, which may be and often are incompatible with those of other members of the society. For instance, the tendency to have total freedom of action for one individual is incompatible with that of other individuals.

In order for a particular society to survive and thereby to secure the survival of its members and of future generations, the inevitable conflicts between individuals need to be resolved as far as possible and in the best way possible. This implies some limitation of the personal freedom of all the people involved. This, in turn, leads to the necessity of establishing rules and laws for living together in the same society, and demands, too, the formation of special institutions to ensure the observation of those rules and laws, and to prevent concrete persons from breaking them. Additionally, the reasonable (if possible) satisfaction of the variety of needs of the individuals demands the appropriate distribution of different kinds of duties among them.

All of this leads to the necessity for any society, no matter how large, to have a particular social structure. Thus, a society, being a set of concrete people, should necessarily consist of classes (see Section 2.3) determined by the social status of the individuals they comprise. This is a natural law for the existence of *any* society of concrete human beings.

It is a well known fact, by the way, that to have a certain structure is a law not only for human societies, but also for any animal societies.

In any real society, the distribution of people over the values of moral qualities is such that most individuals care for themselves and their kinsfolk in the first instance, even if to the prejudice of other people. This is at least one of the reasons why any real society fails to satisfy the personal needs and interests of each and every member in what to the individual seems to be an appropriate way. Dissatisfaction with the society a person lives in, at least by some of its members, is also a natural law of the existence of *any* human society. We thus approach the problem of whether any particular society could ever be considered the best one.

4.5. Social sciences, utopias, and doctrines

Any society forms a finite set of concrete human beings who differ in their individual characteristics both intrinsically and extrinsically, as for instance in their social status. The number of concrete members in a society necessarily changes with time. Its structure as well as its rules and laws may also change. However, the necessity of having some structure remains the invariable law for any society.

As already mentioned (see also Section 2.3), every society may be divided into the classes of concrete people it comprises, classes that correspond to the values of any of their properties. From the standpoint of sociology, it is mainly the social properties of people and, hence, classes corresponding to their values, which are of special interest. The properties meant are those determining the social status of concrete members of the society, namely, relation to the means of production, means of distribution, government, legislation, education, etc., etc.

The necessity for any society to have some structure and to be divided into classes leads to the following important conclusion: *a classless society cannot exist in principle.*

Different people of any class have their own real (or supposed) needs, own interests, own understanding of what freedom is, own feelings and emotions, own desires, and own ideas (if any) about all the things in the world. Also, in any society there are variable distributions of individuals in every class over the values of all their other properties, and in particular over the values of 'intellectuality', as well as 'honesty', 'goodness', and other moral characteristics.

Since the different interests of different concrete people are often incompatible and clash with each other, many try to improve their own situation to the detriment of other members of the society. At the same time, for various reasons, different members of a real society are never equally satisfied with their own (and others') actual conditions and, hence, at least some of them are dissatisfied with the society itself. This fact, which was stated a few pages ago as a natural law of the existence of any society, leads to one more important conclusion: *there cannot exist in principle an 'ideal society', that is, one that might be equally satisfactory for all of its members.*

Some readers may dislike or not believe both conclusions, but they follow inevitably from a consideration of the peculiarities of real, actually existing, human beings, and I do not think they may in any way be ignored.

Throughout all historical periods, the grief and suffering of many poor, miserable, and oppressed people together with the dissatisfaction of many others with their own living conditions or with those of other people made them all desire and dream of having a better society, at least for future generations. This has created auspicious conditions for the widespread mass belief in the feasibility of an 'ideal society' having a perfect political and social system.

Some of the ideas of this kind may take the form of an undisguised utopia. The very word utopia, coined by Thomas Moore from the Greek, means a place which does not exist anywhere on earth, and it goes without saying that the idea of a utopia is only a vain day-dream, a pleasant fairy tale which has nothing to do with reality. Most people understand this and do not take the utopian scheme too seriously.

On the other hand, there are some ideas which pretend to be scientific but are actually latently utopian. One of the related doctrines, namely, that of Marx–Lenin, states that the future of mankind is in a classless society, where people are all equal, and the production of goods is sufficient to permit realization of the slogan: 'from each according to his ability, to each according to his needs'.

Unfortunately, many people blindly believe the doctrine and, even worse, actively try to attain the 'bright future' as soon as possible. It is out of place to discuss here how potentially dangerous for the very existence of mankind are attempts to translate that doctrine into practice. However, to show that its premises are wrong and bear no relation to science may be of significance, the more so because practice has already disproved the doctrine.

As distinguished from natural sciences, where theories can and must be verified by experiment, in social sciences any verification of certain theories is impossible for obvious reasons: no specific social experiments intended to verify a social theory can be performed on sets of living human beings. The only available experimental results which may be used for verifying any such theory are those provided by the experience of mankind, that is, by the history of, and the contemporary developments in, all the known societies. Thus, a social science should be based on the practice of the real life of mankind for the whole of its history.

Practice confirms, however, the aforementioned conclusions concerning the impossibility of the existence of an 'ideal society', in particular a classless society, which have been drawn from a general consideration of the peculiarities of concrete human beings and of the possible relationship of the latter with a society. Yet the doctrines maintaining the feasibility of a classless society are based on the introduction of imaginary people who should live in such an invented, imaginary society. The people invented are all equally wise, honest, and good, and no one has bad moral qualities or low intellectual faculties. Because they are not based on the real properties of concrete human beings, these doctrines are inherently incompatible with science and have never been verified in human experience.

Another doctrine, related to the former ones, claims that in order to construct the classless society, it is necessary, first of all, that the proletariat, that is, the working class, seize power. But no *class* can seize power; only concrete people can do this. Moreover, concrete proletarians, if they seize power, will by definition become members of the ruling class (from which they would never want to return, as practice shows). Thus, this doctrine also is wrong.

In connection with this conclusion it is necessary to stress an important lesson of human history: any attempt to translate into practice the idea of the superiority of one class over others, no matter of which kind it might be (even the 'supreme' race), leads to the formation of societies ruled by totalitarian regimes, which protect their systems by brutal oppression and even annihilation of huge numbers of human beings, while announcing that they are doing it in the interests of the 'people'.

The ultimate practical consequence of treating abstractions as real is the annihilation of huge numbers of individuals in the name of the abstraction. Thus, real people were obliterated in the name of the forth-coming new Soviet Man. But who cared about living people, about individuals? J. V. Stalin, as he named himself, the true follower of Marxist–Leninist ideas was one of the most monstrous and bloody tyrants in the whole of world history, and exterminated many millions of people. He mentioned more than once that the death of one individual is a tragedy, whereas the death of thousands (more exactly, of millions) of people is just statistics.

What is also apalling is that one of Štalin's admirers in the West, the renowned George Bernard Shaw, a brilliant master of paradoxes, bluntly expressed the same idea in one of his works.[8] He wrote: 'The planners of the Soviet State have no time to bother about moribund questions; for they are confronted with the new and overwhelming necessity for exterminating the peasants, who still exist in formidable numbers. The notion that a civilized state can be made out of any sort of human material is one of our old Radical delusions. As to building Communism with such trash as the capitalist system produces, it is out of the question.'

He thus also regarded living people only as a *material* for building such an imaginary 'ideal society'.

Societies of this kind are quickly cast aside when considering the problem of the best possible society, since they really are the worst in all respects. Most members of such societies have been well aware of the harshness, having felt it on their own backs.

The problem of the best society cannot, however, have any general theoretical solution either, since the concept of the 'best society' is obviously not definable in any objective terms independent of any relation to the individual members of the society. Thus, instead, some real practical problems arise; for instance, how can the existing living conditions of the members of a given society be improved, and by how much?

Should this particular problem be set up as a theoretical one, it would be insoluble, too, because its formulation is rather vague; there is nothing in it suggesting who it relates to—does it concern all the members of a society or must criminals, for instance, be excluded? Fortunately, however, the problem is a practical one, and even though it cannot be formulated in exact objective terms, it can be solved in practice, though differently in different particular cases. To be sure, the solution to this practical problem cannot be achieved quickly, the more so because every society is dynamic, and new developments in science and technology as well as international contacts often greatly influence the conditions under which a society develops, and thereby also the conditions under which its concrete members live. Thus, the solution to the practical problem, in every concrete case, seems to be necessarily a particular gradual process of improvement (with possible fluctuations), but it can be achieved eventually. This optimistic conclusion can be justified by the experience of some western countries, even though none of them has yet reached a satisfactory level of improvement.

A number of complicated problems of different kinds arise along with the development of a society, such as moral, psychological, legal, social, political, and, of course, environmental ones. Hence, many pertinent humanistic and natural sciences should be involved in the process of the solution of the main problem—the practical improvement of people's living conditions. When solving any problem of the particular kind just mentioned, however, it is necessary to keep in mind that even the very statement of the problem should be related to concrete people only, and not to any invented or abstract ones. A clear understanding of this fact might help humanitarian scientists very much in finding real ways of helping real people.

4.6 How reliable is a poll?

It is time now to return to probabilities. But before leaving the people-related matters touched upon in the preceding sections, let us take a parting glance at them by mentioning just one characteristic trait of experimental statistics as applied to the examination of some particular problems. This will give us, too, grounds for a more or less smooth transition from the extremely troubled and exciting common problems of mankind to the relatively calm realm of the natural sciences.

Unlike general sociological theories that cannot be tested in principle (see Section 4.5), some aspects of human activity require statistical experimentation: in particular, the gathering of certain statistical data (see Section 2.1). For instance, to get the characteristics of the state of a country, say, in relation to its population, it is necessary to collect the appropriate statistical data. This is usually done by interviewing people inhabiting the country (population census). The method of questioning people (potential voters) is usually used for forecasting the results of an election. The results of a poll carried out on a relatively small group of possible voters can be regarded as yielding an estimate of the probability of the result of the event itself. But how reliable are those results?

To answer this question we should first remember that when performing a statistical experiment on people, one must be very careful as to the objectivity of the statistical data obtained. If the method applied is based on making use of some devices which measure the values of some quantities, for instance, the physical properties of people, then the objectivity of the results can be considered to be secure. However, if the statistical data are collected by questioning people or on the basis of the personal observations of the people who collect statistical material, then in many cases the data obtained are far from being objective and, hence, correct.

In some cases they can hardly be correct in principle. That is the case, for instance, when the people questioned are under the pressure and surveillance of those governing them and who, for certain reasons, do not want the actual facts to be revealed. In these cases the statistical data drawn by the inquirer should be regarded at least as doubtful. This is particularly true when the inquirers do not command the language of the country from which they seek data. Nor are the inquirers or observers always intelligent or honest enough (do not forget that they are individuals with their own psychological peculiarities!); as a result, they may release their flawed data as genuine. It is necessary for everyone to be aware of such a possibility in order not to swallow the bait.

The truth of this can be seen, for instance, by comparing the forecast and actual results of the elections in the USA and in Nicaragua. In the USA, where concrete people can reveal freely whom they would like or not like to be the leader, the actual results of elections usually coincide satisfactorily with the predicted ones. In Nicaragua's first free presidential election of February 1990, the actual result of the election turned out to be quite the opposite of that predicted by polls. The reason for that is fully understandable: those questioned were afraid to voice their real opinions as long as Ortega was in power.

The lesson to be drawn from those facts is clear enough. The 'probabilities' figured out from a poll not infrequently can be different from the true probabilities reflecting the real state of affairs.

Chapter 5
Probability as a base for quantum physics

5.1 Probabilities revisited

In Section 2.1, an explicit definition of probability was given for the particular case of balls in a container. The definition reads: by the probability that a ball in a container belongs to a definite class is meant the ratio of the number of balls in this class to the total number of balls in the container.

A similar definition can be given in other cases involving sets of concrete objects of various kinds, which are available at once. We should note, however, that, even though the wording of the above definition sounds wholesome enough and seems quite understandable, it is, as a matter of fact, somewhat vague. Indeed, the word 'ball' occurs three times there: the first time it is related to an abstract object (a ball in a container), whereas the two other occurrences are connected with concrete objects— the number of balls in the class and the total number of balls in the container. Besides, the expression 'a ball in the container belongs to a definite class' is meant only to indicate that the values of the properties of the abstract ball are determined by that particular class of concrete balls (see Section 2.5).

To simplify the wording we replace the words 'number of balls in the class' and 'total number of balls in the container' by 'measure of the subset' and '(total) measure of the set', respectively, which are valid for the most general case (the term 'class' is replaced by 'subset', the former being a particular case of the latter). Now, keeping in mind the above remarks, we can put the explicit definition of probability as follows (for the cases when a set of concrete objects is available at once).

By the probability that the abstract object corresponding to a set of concrete objects has values of properties which are determined by a definite subset of the set, is meant the ratio of the measure of the subset to the measure of the set.

A few explanatory words about the term 'measure' are in order. This term as applied to sets of concrete objects means the *amount* of concrete objects in the extended sense of this word, that is, the *number* of concrete objects if they can be counted, the *length* of the segment of the line the concrete objects occupy if they are continuously distributed along it, the *area* of the surface the concrete objects continuously lie on, or the *volume* of the space the concrete objects continuously occupy.

By way of example, let us consider the following probabilistic problem. What is the probability that a randomly selected point in a circle of radius

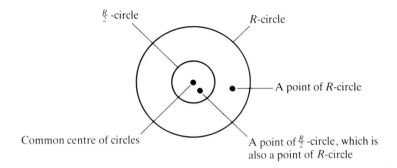

Figure 8. *Concrete points of the R-circle occupy the entire R-circle, and concrete points of R/2-circle occupy the entire R/2-circle.*

R (R-circle) happens to be in the concentric circle of radius $R/2$ ($R/2$-circle)? This problem is equivalent to the problem: what is the probability that a point of the R-circle is one of the $R/2$-circle?

The set of concrete objects in this example is that of all concrete points of the R-circle (see Figure 8). The property of a concrete point of interest to us is its distance from the common centre of the circles. The values of this property for the abstract object 'a point of the R-circle', determined by the $R/2$-circle subset of concrete points, are contained between 0 and $R/2$. Hence, according to the definition, the probability sought equals the ratio of the measure (area) of the $R/2$-circle to the measure (area) of the R-circle, which is 1/4.

It follows from the above definition that probability is a positive number not exceeding unity. All the conventional rules of calculating probabilities also follow directly from that definition. The definition is valid, however, only for those cases where the set of concrete objects is given wholly and is measurable. In most cases, the set of concrete objects is not given wholly. When performing a statistical experiment, the concrete elements of such a set appear at random one by one. That is the case, for instance, in Aspect's experiment (see Section 3.5), in a diffraction experiment, and in many other physical experiments, as well as in raffles and other games of chance.

The set of concrete chords in a circle is an example of a set of concrete objects given wholly, but not measurable (see Figure 9). What can one do in such cases? Is it possible to give an explicit definition of probability for such sets of concrete objects?

It has turned out that the above definition of probability can be extended to the general case of any set of concrete objects which can be subjected to random tests. The only change in the definition is that the measures indicated there should be related to an adequate set (the explanation of this term will follow). With this change only, the explicit definition of probability reads: by the probability that the abstract object corresponding to a set of concrete objects has values of properties which are determined by a definite subset of the set is meant the ratio of the measure of the equivalent subset of an adequate set to the measure of the latter. This generalization of the definition of probability changes nothing in the properties and rules of handling probabilities.

Now, what is an adequate set? This concept is tightly connected to the statistical experiment, that is, to a very long series of random tests performed on a set of concrete objects. Suppose we have two sets of concrete objects, such that to every subset of one set there corresponds a subset of the other set—these are then called equivalent subsets. Suppose, now, that the statistical experiment is performed on both sets. If the frequencies (statistical probabilities; see Section 2.1) of events related to equivalent subsets are approximately equal for all the subsets, then the sets themselves are said to be adequate (or, more strictly, statistically adequate). Certainly, every statistical experiment is performed on some definite set of concrete objects. However, in calculating the probabilities of different events—of the results of the statistical experiment—it does not matter which one of the sets adequate to the tested one is chosen, for they all should give, at least approximately, the same numbers. Let us consider two simple illustrative examples.

In Example 1 of Section 2.1.1, we dealt with a set of 20 balls. It was divided into four classes, BT, RT, BL, and RL, which consisted of 8, 4, 3, and 5 balls, respectively. If we replace this se by a set of 2000 concrete objects of any other shape, which has four classes consisting of 800, 400, 300, and 500 concrete objects, respectively, then this new set will obviously be adequate to the initial one (and vice versa).

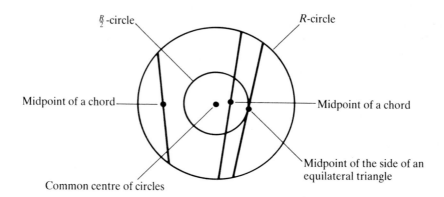

Figure 9. *The midpoint of every concrete chord is a concrete point in the R-circle. The lengths of the chords whose midpoints are in the R/2-circle exceed the side of an equilateral triangle inscribed in the R-circle.*

The set of concrete chords in a circle was mentioned before as an example of a set of concrete objects given wholly, but not measurable. At the same time, the midpoint of a concrete chord is a concrete point within the circle, and to each such point (except the centre of the circle) there corresponds only one concrete chord. When performing the statistical experiment revealing at random concrete points of the circle, one obtains simultaneously the corresponding concrete chords whose midpoints are the former. Hence, the set of concrete points of a circle is adequate to the set of concrete chords therein (Figure 9). This fact makes it possible to settle easily the Bertrand paradox (see Section 2.1).[1, 4] Indeed, the set of concrete chords whose lengths exceed the side of an equilateral triangle inscribed in the R-circle is equivalent to the set of concrete points of the $R/2$-circle, while the set of all the concrete chords of the R-circle is adequate to the set of all the concrete points of that circle. Hence, Bertrand's problem has a unique answer: The probability sought equals the ratio of the area of the $R/2$-circle to that of the R-circle, that is, 1/4.

The concept of the adequacy of sets is a very important one. The problem of calculating the probabilities can be solved if a proper set adequate to the set of concrete objects subjected to random tests is found. Many probabilistic problems have been solved, as a matter of fact, by making use (though implicitly) of adequate sets. Classical statistical considerations, in particular, are actually based on the utilization of adequate sets.

Thus, the first step in solving ordinary (classical) probabilistic problems is the construction of a set adequate to the set of concrete objects under examination. Some readers may wonder if there is such a set. The answer is given by the following substantiated assertion:[1,4] for any set of concrete objects which can be subjected to random tests there exists at least one set adequate to it.

The wording of the basic phenomenon of probabilistics, given in Section 2.1 for the example of balls in a container, can be generalized as follows. When a very large number of random tests have been carried out on a set of concrete objects, the frequency of the appearance of a concrete object of each subset of the set, that is, the statistical probability of this event, is approximately equal to the relative measure of this subset.

This assertion, substantiated by the huge quantity of statistical material gathered during the long history of the existence of experimental statistics, plus the preceding assertion concerning the existence of at least one adequate set for every set of concrete objects capable of being subjected to random tests, plus the definition of probability, together constitute the basis of probabilistics (see Section 2.1). They reflect the connection between the two parts of probabilistics—probability theory and experimental statistics—all the conventional properties and rules of calculating probabilities and related quantities follow directly from the definition of probability.

5.2 States and beyond

The *classical method*, presented in the preceding section, whose first step is the construction of a proper adequate set, is applicable to conventional probabilistic problems and in particular those related to physical macrosystems. It is inapplicable, however, as experiments show, to probabilistic problems related to physical microsystems. To get to the method appropriate for solving those problems, we now move a little deeper into probabilistics.

The concepts to be used in what follows are, in the first place, 'state', 'random variable', and 'mathematical expectation'.

The term 'state' is widespread in many fields of human activity and also in common life. One may speak, for instance, of a state of affairs, or of the economic state of a country, or of the state of a physical system. Although this term cannot be defined exactly by reference to some simpler concepts, its meaning is, as a rule, clear enough and understandable to those who use it. When speaking of a state, we always mean some qualitative or quantitative set of characteristics of an object. One may, for instance, say that the state of health of Mr. N is good (bad or satisfactory). The same qualitative estimates may be used for the characterization of the state of crops, or of the economy, etc.

It is possible to give a quantitative characteristic of Mr. N's health, using numbers by means of which different indices of health are estimated in medicine—number of heart beats per minute (pulse), blood pressure, content of haemoglobin in the blood, etc. For crops and other objects, for which a qualitative estimation of state is usual, a quantitative one is possible too. But for physical systems only the quantitative characterization of a state makes sense. For instance, an equilibrium state of a gas is characterized by its temperature, pressure, and specific volume. The state of a concrete classical mechanical system consisting of several particles is characterized by their coordinates and momenta.

The term 'state' is closely connected with the term 'statistics', and the latter derives from the Latin status. Originally, statistics applied to the use of masses of factual data to describe the state of a country, which includes details about the population given by data obtained by interviewing concrete people inhabiting the country. These statistical data are classified according to different features (properties) of the country's inhabitants (sex, age, education, etc.), and are presented in the form of a summary indicating the distribution of the population over the values of these properties, that is, the absolute or relative number of people (the absolute or relative measure of a subset of people) belonging to each category accounted for in the country's population. Classification of the initial statistical data permits the abstraction of information about the individual peculiarities of the persons interviewed and yields a description of the state of the country's population (or, in other words, the state of an abstract inhabitant of the country) with the aid of the relative measure of subsets of people belonging to different categories of the population.

However, by definition (see the preceding section) the relative measure of a subset of concrete objects is the probability that the pertinent abstract object has the corresponding values of the properties of these concrete objects. Therefore, classification of the statistical material leads, in the case under consideration, to a description of the state of an abstract inhabitant of the country by indicating the probability distribution for him to have different values of the properties of the concrete inhabitants.

It might seem that the term 'state' should refer to the whole set of concrete objects under examination. First, however, as noted before, the set is not always given wholly, and not always measurable. Second, characterization of a state of a set of concrete objects by means of the relative measure of its subsets corresponding to different values of the properties of the concrete objects coincides with the indication of the distribution of the probability for the corresponding abstract object to have the respective values of these properties. Therefore, in probabilistics we will regard the term 'state' as referring to abstract objects and suppose that a state of an abstract object (it is also possible to say: 'a state an abstract object is in') is characterized by the probability distribution for it to have various values of the properties of the concrete objects corresponding to it. This makes sense for any sets of concrete objects which are capable of being subjected to random tests, even in those cases where they are not given wholly or are not measurable.

A related important concept in probabilistics is that of a 'random variable'. A random variable is defined in probabilistics simply as a *property* of concrete objects (belonging to a definite set of them) whose values (perhaps in proper units) are *real* (that is ordinary) *numbers*. This definition differs from the more complicated one used in ordinary probability theory, but it is fully in line with the usual wording. Thus, for instance, the angle of deviation from the vertical of the thread of a pendulum, the coordinates and momenta of a particle, the spin of a particle along a chosen axis, the polarization of a photon with respect to a certain p-line (see Section 3.4), etc., are all *random variables*.

'But how can such properties be called random variables?', one may ask; 'every concrete object should have one and only one value of each of its properties, as was explained in Section 2.2. What, then, is random about them?' The answer is this. It is true that every concrete object has one and only one value of each of its properties, but in performing random tests on a set of concrete objects the values of the properties also are revealed at random. Hence the 'randomness' of a random variable manifests itself only in random tests.

Properties of many concrete objects are random variables, that is, they have appropriate numerical values by their very nature. An important point, however, is that properties of any concrete objects can be represented conditionally by random variables. Such a representation can be achieved in an infinite number of ways, and the only requirement to be satisfied is that different numbers should be assigned to different values of a property. This is necessary in order to distinguish between different concrete objects.

In Example 1 of Section 2.1.1, for instance, we can ascribe any two different numbers to the colours 'red' and 'blue' of the balls in the container, and any two different numbers to the marks 'L' and 'T' on them. If we chose 1 for 'blue', 2 for 'red', 1 for 'L', and 2 for 'T', the classes into which the balls in the container are divided will become: 11, 12, 21, and 22. Any other assignment of different numbers to the colours of and marks on the balls in the container is equally good.

We have already used this way of representing properties of concrete objects in the case of the polarization of photons (Section 3.4). In fact, we have ascribed a random variable to the polarization of a concrete photon with respect to a chosen p-line, with two values: 1 (for parallel polarization) and -1 (for perpendicular polarization). The spin component along an axis of a concrete spin-$\frac{1}{2}$ particle (in terms of /2 , where is the Planck constant h divided by 2π) is a random variable with two values, 1 and -1, also. In Section 1.4, a computer was mentioned that can select at random, with equal probabilities, one of two numbers, 0 or 1. The number being selected is also a random variable whose values are 0 and 1. We can introduce a random variable with two values 1 and -1 for a coin which is being tossed by ascribing, for instance, the value 1 to the head, and -1 to the tail of a concrete coin. We can equally well use a random variable with two values 1 and -1 for Schrödinger's cat, by ascribing, for instance, the value 1 to a live cat and -1 to a dead cat.

It is interesting to note that in all the above examples the probability distributions over the values of each random variable are the same (1/2 for one and 1/2 for the other) even though the random variables themselves are of different origin and nature: for photon polarization, coin tossing, and Schrödinger's cat, the respective random variables are introduced artificially, whereas for spin components and numbers chosen by a computer they arise naturally; additionally, for photon polarization and spin components, the random variables are related to physics, whereas in the cases of coin tossing, Schrödinger's cat, and numbers chosen by a computer they have nothing whatsoever to do with physics.

As can be seen from the above, the introduction of random variables makes it possible to treat different probabilistic problems from one and the same unified point of view. Now a state of an abstract object can be characterized by the probability distributions over the values of all the random variables involved or, in other words, by the probability distributions of those random variables. One more point should be stressed, by the way, which also follows from the above. Random variables and probability distributions are altogether different things. The probability distributions for a state of an abstract object are determined mainly by the objective composition (structure) of the set of concrete objects to which the former corresponds; the values of random variables are in a way arbitrary and depend on how the latter are chosen.

One more important related concept is that of the 'mathematical expectation' of a random variable. This is a theoretical quantity of probability theory. Its counterpart in experimental statistics is the arithmetic average of a random variable, that is, the algebraic sum of the products of the values of the random variable by the frequencies (statistical probabilities) of their appearance. The (statistical) arithmetic average is widely used in everyday life, for instance, in economics, the insurance industry, etc., and this concept should not need, I believe, any further comment here. Since the values of random variables are the same in both experimental statistics and probability theory, and probability is the counterpart of frequency in the latter, the mathematical expectation of a random variable has the same algebraic form as the arithmetic average: the former is obtained from the latter by simply replacing frequencies (statistical probabilities) by (theoretical) probabilities. In the case of continuous random variables, such as coordinates of particles, for instance, the probability should be replaced by the probability density (that is, probability per unit of length, area, or volume), and summation by integration, but this is only a mathematical trick which changes nothing in the essence of the matter.

The mathematical expectation of a random variable can be represented in different mathematical forms, and this in turn leads to different ways of describing a state of an abstract object. The concept of a 'description of a state' as now introduced is wider than that of a 'characteristic of a state' which had been used before. The latter has a direct probabilistic meaning—the characteristic of a state is given by the probability distributions in the state for all the random variables involved. The former is given by any auxiliary mathematical quantity which determines uniquely the pertinent probability distribution, but may not have a direct probabilistic meaning. Different ways of representing the mathematical expectation result in different mathematical methods of solving probabilistic problems. In particular, the mathematical technique arising from one specific mode of the representation of mathematical expectation happens to be the one conventionally used in quantum physics. It belongs, however, to probabilistics itself, and its application in quantum physics is due only to the fact that quantum physics deals just with the probabilistic problems concerning physical microsystems.

This specific method of probabilistics has been called by me the 'quantum approach'. Unlike the classical method touched upon before, which is valid for physical macrosystems but not for microsystems, the quantum approach is valid for both. In view of my promise to avoid mathematics, I may only drop a hint about how a specific way of representing the mathematical expectation of a random variable results in the quantum approach of probabilistics.

In each of the above cases, except for the computer related one, the random variable has two values, 1 and -1, with equal probabilities, 1/2 and 1/2 each. Hence, for each of these cases, the mathematical expectation is

$$1/2 \times 1 + 1/2 \times (-1) .$$

Let us now represent this mathematical expectation as

$$(1/2)^{1/2} \times 1 \times (1/2)^{1/2} + (1/2)^{1/2} \times (-1) \times (1/2)^{1/2}$$

Now, instead of a pair (1/2, 1/2) of probabilities which characterize the state of each corresponding abstract object (in particular, the state of an abstract photon but also that of an abstract Schrödinger's cat) we have a pair $[(1/2)^{1/2}, (1/2)^{1/2}]$ of numbers which are the square roots of those probabilities. This pair of numbers, which represents some mathematical vector (that is, a mathematical generalization of conventional vectors), does not have a direct probabilistic meaning, but the pair does determine uniquely the corresponding probability distribution (for the squares of these numbers are the respective probabilities) and, hence, suits well for the description of the relevant state. It suffices now to find the vector *describing* the state of interest, in order to get the corresponding probability distributions.

The above record of the mathematical expectation is the simplest one. Its more complicated modification leads to the introduction of special mathematical quantities (matrices and operators) for the representation of random variables, and multidimensional (mathematical) vectors and functions for the description of states of abstract objects. And that is the way the quantum approach arises. Thus, step by step, and without any dramatics, we have reached the realm of quantum physics. Let us now intrepidly enter it, with our eyes open, leaving behind any possible prejudices and misconceptions, and let us try to understand and learn as much as possible, at least qualitatively (for we are not introducing mathematics).

5.3. What is quantum physics?

Quantum physics is the science that deals with the theoretical (probabilistic) and experimental (statistical) study of physical *microsystems*. The specificity of the probabilistic behaviour of these systems is reflected in the specific character of the mathematical methods that *must* be used for the description of that behaviour and the solution of the related problems. Unlike mechanical macrosystems, whose probabilistic characteristic is determined by the laws of classical mechanics valid for every concrete system,[4] the origin of the probabilistic behaviour of microsystems is unknown, at least so far. Will it ever become known? I do not think that anyone can answer this question at present.

Nonetheless, even though we know nothing about the laws governing the behaviour of concrete microsystems, we are well equipped with the mathematical methods, originating from the quantum approach of probabilistics, which allow one to find the related probabilistic characteristics of the abstract microsystems of interest.

Quantum physics may also be said to be the application of probabilistics to physical microsystems. Hence, quantum physics is a domain of *probabilistic physics* which is the application of probabilistics to physics in general. Another domain of probabilistic physics is *classical statistical mechanics*, which is the application of probabilistics to mechanical macrosystems.

In probabilistic physics, including quantum physics and classical statistical mechanics, concrete objects and abstract objects are concrete physical systems and abstract physical systems, respectively. Properties of objects are, in particular, various physical quantities, including coordinates and time. Since the values of these properties (in proper units) are real numbers, physical quantities prove to be random variables by definition.

One of the main tasks of probabilistic physics is to find the probability distributions of certain physical quantities and to calculate the mathematical expectations of some of their functions in states of an abstract physical system. The problem for microsystems can be solved, as was mentioned earlier, by the application of the quantum approach only, which leads to the mathematical techniques being used in conventional quantum physics.

It is worth emphasizing that the meaning of probability and the rules it obeys, in particular, rules of addition and multiplication, are the same, regardless of whether the classical method or the quantum approach has been utilized, as has been demonstrated elsewhere.[4] The term 'quantum probability', often used in conventional quantum physics, should be regarded, therefore, as reflecting only the way the probabilities have been calculated, that is, by making use of the quantum approach.

The first step in solving a particular problem is finding the state vectors describing the states of the abstract system under examination. This is a purely mathematical problem which consists of the construction and solution of some particular equations. However, I am not going to present any equations: my only intention is to explain the origin and meaning of some of them and, in particular, of the famous Schrödinger equation.

We shall consider, first, some simple cases. Suppose there is a set of concrete objects each of which generally has its own value of a certain property they all possess. The state of the abstract object corresponding to the set can be described by a state vector (state function) which determines uniquely the probability distribution of that property (random variable), that is, the distribution of the elements of the set over the values of the random variable. The mathematical expectation of the random variable in this state generally differs from the values the concrete elements of the set possess.

It can happen, however, that all the concrete objects of the set have one and the same value of that random variable. In this case the mathematical expectation of the random variable coincides with that value, which is the only one the abstract object has in this state. The requirement that an abstract object had a definite value of a certain random variable leads to an equation whose solution is a vector (state function) describing just this state of the abstract object.

A set of concrete free particles of the same mass moving along, say, the x axis of some (inertial) coordinate system at one and the same rate is a good example of the above case. Different concrete free particles *must* have different x coordinates, but one and the same momentum. The requirement that the abstract free particle has a definite value of momentum—the same as the concrete particles have—leads to an equation whose solution is a function of the coordinate x describing this particular state of the abstract particle. The magnitude squared of this function, which is proportional to the probability density of the coordinate x, is constant. This means that the probability of finding a free particle in some interval along the x axis is proportional to the length of the interval. So, in this state of the abstract free particle the value of the momentum is definite, whereas the value of its x coordinate is uncertain. It is interesting to note that the same result follows from the application of the classical method to this case (see Figure 10).

Here is another example. Every element of a set of concrete spin-$\frac{1}{2}$ particles of one kind has one and the same value 1 of the spin component along, say, the z axis (see Section 3.2). The requirement that the abstract particle has a definite value of that spin component (the same as the concrete particles have) leads to a simple equation whose solution describes the relevant state of the abstract particle.

Suppose now that some other axis (b axis) is chosen, which makes an angle d with the z axis. What can be said about the value of the spin component along the b axis of a concrete particle belonging to the set? The answer is 'nothing'. Every concrete element of the set has the definite value 1 of the spin component along the z axis. However, no spin component along any other axis can exist simultaneously (see the end of Section 3.2) and, hence, its value cannot be predetermined by the original value and the

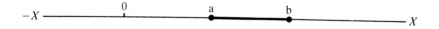

Figure 10. *The set of concrete points on the x axis is adequate for the set of identical concrete free particles moving along the x axis at one and the same rate. The probability for the corresponding abstract free particle to have its x coordinate within an interval ab is proportional to the length of the interval.*

choice of the b axis. At the same time, an equation can be constructed whose solution describes the state of the abstract spin-$\frac{1}{2}$ particle with respect to a spin component along any b axis, and thereby determines the probability of that spin component having the value 1 or −1. The same can be done for the polarization of photons (see Section 3.4).

The above examples belong to different parts of quantum physics. The first concerns a mechanical system, a free particle, and has an analogy in classical mechanics. The second has no analogy in classical mechanics. Quantum physics comprises, in general, a diversity of microsystems, such as particles, atoms, molecules, crystals, etc. Some of them have classical mechanical analogies, and some do not.

The probabilistic characteristic of a microsystem that has an analogy in classical mechanics is described by a solution of the appropriate Schrödinger equation. This equation has proved to be a particular case of the general equation in probabilistics (related to a special class of random variables that satisfy the so-called Hamiltonian equations).[4, 9] Before proceeding to the Schrödinger equation I would like, however, to conclude this section with a few words on the terminology related to the field we are dealing with.

The relevant books and articles, both scientific and popular, refer to seemingly different things: wave mechanics, matrix mechanics, quantum mechanics, quantum physics, and quantum theory. Are these different terms related to different branches of science? The answer is no: they all actually concern one and the same thing, namely, the probabilistic treatment of physical microsystems. Different names for really the theoretical part of the branch of science under consideration reflect the history of its development.

Wave mechanics was suggested by Schrödinger (1926) who proceeded from the conviction that, since there is an analogy between geometrical optics and classical mechanics, a similar analogy should exist between wave optics and the supposed wave mechanics. Using this analogy, he derived the basic equation of wave mechanics—the Schrödinger equation.

Matrix mechanics was introduced by Heisenberg (1925), proceeding from quite a formal consideration of atomic spectra, in which the technique of matrix calculus was utilized. However, it soon turned out that the mathematical techniques of wave mechanics and matrix mechanics were only different representations of the general techniques of linear algebra, a special branch of mathematics.

The terms quantum mechanics, quantum physics, and quantum theory appeared later and actually supplanted the first two. Out of all those terms I prefer 'quantum physics', because it reflects the facts that it is a whole science, not only a theory, and concerns physics in general, not only mechanics.

In the rest of the book we will consider some selected items of quantum physics and related issues, and we start with the Schrödinger equation.

5.4. The Schrödinger equation: its origin and meaning

As was mentioned before, the Schrödinger equation is related to mechanical systems. Therefore, we should first recall some facts of classical mechanics.

The spatial characteristic of a concrete mechanical system consisting of particles is given by a set of coordinates related to every particle or, in the general case, by a set of so-called generalized coordinates which are functions of the former. The kinematic characteristic of the system is generally provided by a set of an equal number of so-called generalized momenta. Generalized coordinates and generalized momenta are coupled in pairs. This means that to every generalized coordinate the corresponding generalized momentum is assigned by a special mathematical procedure. (In what follows the adjective 'generalized' will be omitted.)

Time differs from spatial coordinates in meaning, but from the mathematical standpoint the time coordinate is equivalent to any spatial coordinate. Hence, there should be the momentum assigned to time, which has proved to be the *energy* with a minus sign of the concrete mechanical system.

Another important physical quantity is the *Hamiltonian function* which is generally a function of all the spatial coordinates, momenta, and time. The real motion of a concrete mechanical system is determined by a set of equations, the Hamiltonian equations, which involve the Hamiltonian function.

It has also been found that any motion of a concrete mechanical system is characterized by the equality of its energy and its Hamiltonian function. The importance of this equality should be emphasized strongly, for the Hamiltonian function and energy are two quite different physical quantities—the former is a definite function of all the spatial coordinates, momenta, and time, whereas the latter is the momentum (with a minus sign) coupled with time. This equality constitutes the necessary condition under which any motion of a concrete mechanical system can occur.

If any readers are unable to recall all the above facts of classical mechanics, rest assured it does not matter. The only important point which should be kept firmly in mind is that every concrete mechanical system possesses two different physical quantities—the Hamiltonian function, and the energy—and that the values of these two quantities must be equal for any motion of the system.

Now a few words should be provided on the classification of mechanical systems. There are three basic types of system: closed, conservative, and non-conservative. A closed system is one which does not interact with any other system and, hence, is not acted upon by any external forces. A conservative system is one which *can* be acted upon by external forces, but the latter do not depend on time. A closed system can be considered a particular case of a conservative system. A system that is acted upon by external forces which depend on time is a non-conservative system. The Hamiltonian function of a conservative system does not depend on time. Hence, neither does its energy. This means that any motion of any concrete conservative system proceeds at a constant value of energy.

The case of a non-conservative system is more complicated, and it may not even be regarded as a separate system, for it is acted upon by forces exerted by external systems, from which the former can never be isolated.

When proceeding to the probabilistic treatment of a corresponding abstract mechanical system, we should recall that all the physical quantities mentioned above, namely, spatial coordinates, momenta, time, energy, and the Hamiltonian function, are random variables by definition. For macrosystems, the application of the classical method (see Section 5.2) leads directly to equilibrium classical statistical mechanics. For microsystems, only the quantum approach is valid.

When making use of this approach, the classical mechanical requirement of the equality of the Hamiltonian function and energy for any motion of concrete systems should be taken into consideration somehow. To do so we recall, first, that an abstract microsystem may not have definite values of energy, but it always has a definite value of the mathematical expectation of energy in each of its states. It should also have a definite value of the mathematical expectation of the Hamiltonian function in every state. So, for an abstract microsystem, instead of the equality of the values of the Hamiltonian function and energy themselves, we may speak only of the equality of the respective mathematical expectations as a necessary condition for the reality of motion of the corresponding concrete microsystems.

Recall next that, when applying the quantum approach (see Section 5.2), the random variables (physical quantities, in the case under consideration) are represented by special mathematical quantities (in particular, by 'operators'); a state is described by a function of some variables (which are the values of pertinent random variables), and the mathematical expectation of a random variable in any given state involves both its appropriate operator and the function describing the state. Now, since the Hamiltonian function and the energy are different physical quantities, the operators for them are also different. Therefore, the equality of their mathematical expectations is an equation with respect to the function describing the state of the abstract microsystem. Why is it an equation? This is because not every function satisfies the requirement of the equality of the mathematical expectations of those two different physical quantities. It transpires that the equation that follows from that requirement is the famous Schrödinger equation.

Thus, the Schrödinger equation stems, in the final analysis, from a specific requirement analogous in a way to that of the equality of the Hamiltonian function and energy for any motion of concrete macrosystems, and its meaning is evident from the fact that the solutions to it describe the real states of the corresponding abstract system, that is, the states corresponding to real motion of the concrete systems which conform to the latter. The probability distributions of the related physical quantities can be obtained from those solutions, and must be in line with the proper statistical experiments if the solutions are correct.

A state in which an abstract system has a definite value of energy is called a 'stationary state'. For such a state, the corresponding concrete systems all have the same value of energy as well. A free particle is an important example of a system of this kind. In every stationary state of an abstract free particle, energy has a definite value, and the same value of energy is possessed by each of the corresponding concrete particles. The Hamiltonian function of a free particle is related to the square of the magnitude of its momentum, which is also constant for a free particle (see Section 5.3). Since the values of the Hamiltonian function and energy must be equal, the same relation holds for the energy of a free particle. This relation has been shown to lead to the wave equation of the same well known form that is valid for any wave processes considered in classical physics (waves on water, sound, optical waves, etc.), which are responsible for such typical wave phenomena as diffraction and interference.[9] The explanation of the origin of these phenomena will be outlined qualitatively in Section 6.1.

As already mentioned, the Schrödinger equation is the basic equation for the probabilistic treatment of a wide range of mechanical microsystems, which includes particles, atoms, and molecules. It was introduced first by Schrödinger himself, proceeding from an analogy between mechanics and optics. Many physicists believe that this equation should be taken for granted as one of the postulates of conventional quantum mechanics. Some (in particular, Dirac and Landau) tried to derive it from theoretical considerations, and they seemingly succeeded in doing so. But their procedures were incorrect, for they mixed up the Hamiltonian function and energy, as has been shown elsewhere.[4]

At the same time, I would like to stress this again, in probabilistic physics, the Schrödinger equation appears simply as a corollary of the requirement analogous to one which holds for any motion of concrete mechanical systems, and its meaning becomes thereby absolutely clear.

I now turn to the often debated issue of the wave—corpuscular duality of particles which, in a way, is also related to the Schrödinger equation.

Chapter 6
Reality versus fiction

6.1 What is a particle—is it a corpuscle, wave, or both?

The above question inevitably arose when the wave properties of particles, suggested by L. de Broglie, had been confirmed experimentally. Particles seem to have, according to the experiments, both corpuscular and wave properties. The former are revealed when particles move in sufficiently definite paths, the latter manifest themselves through such typical wave phenomena as diffraction and interference. One usually concludes from the experiments that both are inherent in every individual particle, and it is precisely this inconceivable-to-common-sense conclusion which causes the bewilderment. How can one and the same thing both move as a particle and propagate as a wave?

It is interesting to note that in the history of physics the corpuscle-or-wave question related to the nature of light arose much earlier. Newton believed that light consisted of particles of extremely small mass. He explained easily, proceeding from this idea, the phenomena of geometrical optics—reflection and refraction—but encountered difficulties when he tried to explain the phenomenon of interference: in particular, the origin of the Newtonian rings he had discovered. For many decades, Newton's corpuscular theory of light was dominant. Then a wave theory of light was introduced by Huygens and developed by Fresnel which also explained the phenomena of geometrical optics, including the rectilinear propagation of light. At the same time, the wave theory had also easily explained the phenomena of interference and diffraction and, by the end of the 19th century, had finally won the competition with the corpuscular theory of light.

A curious fact related to the competition is this. In 1818, when considering a memoire of Fresnel submitted to the Paris Academy, Poisson came to the conclusion that if Fresnel's reasoning is correct then in the very centre of the shadow of an opaque disk illuminated by a point source there must be a point of light. Poisson believed that this seemingly paradoxical result proved the unsoundness of Fresnel's theory. However, an experiment performed by Arago showed that Poisson's conclusions were in line with reality and, hence, confirmed the correctness of Fresnel's theory (see Figure 11).

One more interesting historical fact concerning the corpuscle-or-wave question is worth mentioning. When Roentgen (1895) discovered x-rays, he

Figure 11.   *Diffraction by an opaque disc illuminated by a point source. In the very centre of the shadow of the disc there is a point of light.*

was convinced that they presented beams of some particles, for he could not find any indication of interference or diffraction with those rays. Only much later were the wave properties of x-rays established, and the radiation was recognized as light of very short wavelength.

Thus, the corpuscle-or-wave question was not new in physics and, for ordinary light and x-rays, it was seemingly and ultimately answered in favour of waves, due to the interference and diffraction phenomena they revealed. Later, however, the corpuscular properties of light were found in such phenomena as the photoeffect and the Compton effect, in particular, which led to the consideration of light as consisting of particles (photons), so the corpuscular-or-wave issue reappeared.

The issue could be settled neither in the framework of classical physics nor in that of conventional quantum mechanics, and has provoked long-lasting debates. The pile of works dealing with this and related matters is growing ever larger, involving new ideas, Gedanken-experiments, real experiments, etc., and no end seems to be in sight.

A careful examination of particle diffraction removes the above-mentioned bewilderment. It turns out that there is no such (real) thing that both moves as a particle and propagates as a wave. The so-called 'wave properties' are connected with the probability distributions related to relevant abstract particles, which are revealed (approximately) through the experimental statistical distributions of the corresponding concrete particles, in a long enough series of experiments (random tests). If a few concrete particles only are used in a diffraction experiment, no distinguishable diffraction pattern appears. Hence, only (real) corpuscular properties are inherent in every individual (concrete) particle, whereas the (unreal) 'wave properties' are related to abstract particles. And this fact removes the challenge to common sense.[10]

To substantiate the above statement let us consider in more detail the phenomenon of diffraction and, in particular, that of particles. Diffraction is known to occur when a series of any kind of running waves of the same frequency encounters impermeable obstacles; the diffraction pattern being determined mainly by the diffracting-system-incident-beam geometry and the wavelength of the wave involved, no matter what its nature. For a genuine physical wave process, the diffraction pattern, that is, the spatial intensity distribution, is governed mainly by the corresponding wave equation, which is of the same form for any wave process. Therefore, the diffraction patterns should also be basically alike for any wave processes, even though the nature and type of the waves may be altogether different, as in the cases of acoustic and light waves, for instance. This should be true for the particle diffraction as well, which suggests the question: what is the nature of particle diffraction? If the conventional belief that wave properties are inherent in an individual (concrete) particle is correct, then every concrete particle should have some property which obeys the corresponding wave equation of the usual form. In this case, however, the diffraction pattern must remain the same, regardless of the intensity of the beam, gradually weakening as the beam does. But that is not the case, as experiments show.

The well known assertion that a diffraction pattern does not depend on the particle beam intensity should be understood correctly. That is true only when the total energy diffracted is the same. This means that the weaker the beam the longer should be the duration of the experiment, in order to get the same picture. But, if only a few concrete particles are used, no continuous diffraction pattern appears at all. Instead, some point marks on the display screen can be observed, which are left by the concrete particles involved. This means that no specific wave property is possessed by a concrete particle: no property that might obey the corresponding wave equation. What is more, actually there is no diffraction pattern for particles at all, in the true sense of the concept. When the number of concrete particles is small, the picture obtained is not obviously a diffraction pattern—it is not continuous, and the point marks left by the concrete particles are dispersed on the screen. When the number of the concrete particles involved is large enough, the picture looks like a diffraction pattern but actually it is not, for it still consists of point marks. The experimental made-of-points diffraction pattern may, however, be regarded as an approximation of some continuous one which represents the solution of a certain wave equation involving a certain quantity. This is true for any particles, including photons.

New questions now arise. What is the wave equation in the case of particles? Does it describe some physical process or something else? What is the meaning of the specific quantity obeying that equation?

Before answering the questions, let us dwell a little longer on the diffraction experiments for particles. An experiment performed with a very weak intensity beam should be continued for a long time in order to get a diffraction pattern of desirable brightness. Suppose the experiment is interrupted a large number of times, so that in every time interval only a few concrete particles pass through the diffractometer. If the conditions of the experiment are the same for all the intervals, then the final result will be practically the same as for the continuous experiment. This result does not depend on the duration of the intermissions between the intervals either. If it is possible to record the result for each interval on a separate display screen and then combine them for all the intervals, the pattern obtained will practically coincide with that for a continuous experiment of the same duration, even though on every screen only separate point marks may be found. Further, suppose that the experiment is performed on a separate diffractometer in every interval. If the conditions of the experiment are the same for all the diffractometers, the combined result will also reproduce practically the same diffraction pattern.

These considerations show that a diffraction experiment on particles looks like a typical statistical experiment, in which concrete particles, being subjected to independent random tests, hit different points on the screen by chance; the statistical (relative) frequency of hitting an area in the vicinity of a point is approximately the value of the probability that a particle will do so.

In a particle diffraction experiment we deal with a beam of concrete particles encountering impermeable obstacles, and the experimental statistical distributions of the coordinates of diffracted concrete particles. In the theoretical treatment of particle diffraction we are interested in finding the relevant probability distribution of the coordinates of the corresponding abstract diffracted particle. As was explained in the preceding section, the distribution sought can be found from the solution of the appropriate Schrödinger equation. But what does this equation have to do with the wave equation we are seeking?

The answer comes from the fact that, in a particle diffraction experiment, a concrete particle is free in any part of the space available to it. Hence, the Schrödinger equation should refer to an abstract free particle which is to be in a stationary state with definite values of energy and momentum (see preceding Section). The connection between these two physical quantities makes it possible to convert the Schrödinger equation into a wave equation of the form required, and this allows us to obtain answers to all the above questions.[10]

The wave equation which determines the diffraction of particles is nothing other than a modified Schrödinger equation for an abstract free particle. It does not describe a physical wave process, but is related to the probability distribution of particle coordinates in a stationary state of an abstract free particle. The specific quantity obeying the equation has no direct physical meaning—it is an auxiliary mathematical quantity, called a wavefunction, whose magnitude squared determines the probability density (that is, probability per unit of length, area, or volume) of the particle coordinates.

The famous de Broglie relation defining the 'wavelength' and the Einstein relation defining the 'frequency' of the wave follow immediately from a comparison of the original Schrödinger equation with its modification of the wave equation form.

The widely discussed 'two-slit Gedanken-experiment' (or 'double-slit experiment') is a typical diffraction experiment. Some physicists regard it as a decisive one favouring the assertion that an individual particle does not have a trajectory and behaves like a wave. The debates around this Gedanken-experiment and related issues have not yet quit the stage. It is interesting, therefore, to discuss this Gedanken-experiment in the light of the above considerations, and we will do this next.

6.2 The two-slit Gedanken-experiment: what does it prove?

'Why a thought experiment? Why not a real one?', one may ask. The answer comes from the fact that, in order to get a diffraction pattern from two slits, the distance between the slits should be of the order of a few 'wavelengths' related to a particle. But the 'wavelength' of even a slow electron is about one hundred million times smaller than one centimetre ($\sim 10^{-8}$ cm). Hence, the realization of such an experiment even for slow electrons is practically impossible. Since the 'wavelength' of a particle is inversely proportional to both its mass and velocity, the situation in other cases is no better.

In short, we have no alternative but to consider a Gedanken-experiment. However, no matter whether the experiment is real or in thought, the discussion of its results remains the same.

Of the different versions of the two-slit (double-slit) Gedanken-experiment, we have chosen the following (non-essential details being omitted). A collimated (parallel) beam of free particles all of one kind and of the same energy encounter a screen having two parallel long slits in it (see Figure 12). The particles, after passing through the slits, are recorded on a display screen. If a very large number of particles pass through the slits, they produce a specific pattern on the display screen. The pattern depends on whether only one slit or both have been opened during the experiment. The distribution of particles in the case when both slits are open differs from the simple superposition of the particle distributions which one obtains by opening the slits one at a time. Instead of a simple superposition of the latter distributions, a typical interference pattern appears (see Figure 13). It can even happen that when both slits are open, particles avoid some locations which they used to reach in considerable numbers when one slit alone was open. Usually, this result is discussed in the following manner.

Suppose a particle were moving along a definite path that takes it through one definite slit. Then it would reach a certain point on the display screen after passing through the slit; and it must naturally hit the same point on the display screen (with the same certainty) even if the second slit too is open. The fact that in reality particles stop impinging on the initial spot when both slits are open proves that the supposition of a particle moving

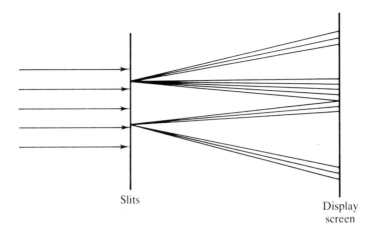

Slits

Display
screen

Figure 12. *Double-slit Gedanken-experiment. A parallel beam of free particles all of one kind and of the same energy encounter a screen having two long parallel slits. The particles that pass through the slits are recorded on the display screen.*

along any path is wrong in the case under consideration and must therefore be discarded. This means, too, that a particle should be considered a wave that passes both slits simultaneously and interferes with itself.

This seemingly correct reasoning proves flimsy, however, upon more careful consideration. Recall that any real experiment deals with concrete particles. Hence, so also does a Gedanken-experiment, and the discussion of its results should imply just concrete particles. Recall, too, that any diffraction experiment, including the one under examination, is an essen-

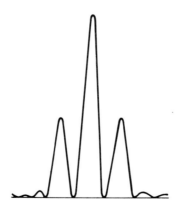

Figure 13. *A double-slit diffraction pattern. The maxima on the curve correspond to the spots most often hit by particles. The minima correspond to the spots which are never hit by particles.*

tially statistical experiment. Now, the conclusion that 'the supposition of a particle moving along any path is wrong in the case under consideration' was drawn from the fact that the supposition of a particle moving along a *definite* path led to a contradiction. However, this supposition may be countered by suggesting that a concrete particle does not move along a *definite* path (rather than the assertion that it does not move in a path at all). This means that the results of the above experiment are fully compatible with the idea that a concrete particle moves in a concrete *random* path, since the only conclusion one may draw from that experiment is, in fact, the following: the totality of real random paths along which a concrete particle can move is determined by the conditions under which the statistical experiment is carried out, that is, by whether the slits are both open or one open at a time. This conclusion is in line with the well known fact that diffraction occurs also in the case when only one slit is open (see Figure 14). In this case, too, a concrete particle does not move along a *definite* path—otherwise no 'diffraction pattern' would appear—it moves along some concrete random path.

The circumstance that the diffraction patterns for one slit and for two slits are different is only a particular case of the general state of affairs. Both the theory and practice of diffraction convincingly prove that every arrangement of a diffraction experiment results in a diffraction pattern specific for it. Different arrangements yield different diffraction patterns. It is precisely this fact which is used widely in many areas of contemporary science and technology, for instance, in revealing the structure of crystals and molecules.

The above discussion demonstrates that the reality of motion of concrete particles along concrete random paths does not contradict the imaginary 'wave' properties of the corresponding abstract particle. The preceding and present sections have already demonstrated that a concrete particle does not have any wave properties. Moreover, there is one more experiment whose results decisively prove this assertion: it is Wheeler's delayed-choice experiment to which we shall now turn.

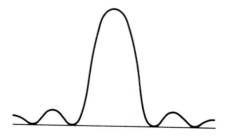

Figure 14. *A one-slit diffraction pattern. In this case, also, there are spots most often hit by particles (maxima on the curve) and ones that are never hit by any particle (minima on the curve).*

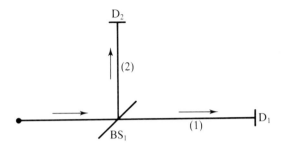

Figure 15. *Open version of Wheeler's experiment. A photon encounters the beam splitter BS_1 and can either pass through or be reflected in a perpendicular direction. Detectors D_1 and D_2 count the numbers of photons passing through the respective routes 1 and 2.*

6.3 Wheeler's delayed-choice experiment

This delayed-choice experiment suggested by Wheeler[11] deals with photons, but its results should be valid for any particles. The outlines of the experiment can be represented schematically as follows.

An appropriate device allows one to perform the experiment in two different versions which will be called, for short, 'open' and 'closed'. In the open version (see Figure 15), a photon encounters a beam splitter (BS) and can either pass through it or be reflected in a perpendicular direction. A detector at the end of each route counts the number of photons that have reached it. If the beam of photons is so weak that the time interval between two different photons encountering the beam splitter is large enough, every individual photon is counted separately and the records of different photons never coincide in time. This version thus displays the corpuscular properties of photons.

In the closed version (see Figure 16), appropriate mirrors on the two routes above cause the photons to change their directions and intersect at a second beam splitter, at which each photon can again either pass through or be reflected in a perpendicular direction. Detectors at the ends of the two final routes count the number of photons that have passed them. It transpires that the ratio of the numbers of photons that have passed both final routes depends on certain details of the performance of the closed version, namely, on the difference between the 'optical paths' (or 'phases') of the routes, which implicates a kind of interference. This version thus reveals the wave properties of photons. The fact that interference occurs when even only one photon is in the apparatus at any time is regarded usually as the evidence that wave properties are possessed by an individual photon (particle) which interferes with itself and, hence, should travel both routes simultaneously.

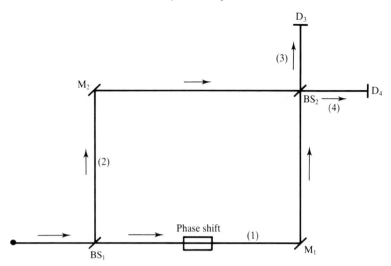

Figure 16. *Closed version of Wheeler's experiment. Mirrors M_1 and M_2 make the photons passing routes 1 and 2 of the open version change their directions and encounter the second beam splitter BS_2. Here each photon can again either pass through or be reflected in a perpendicular direction. Detectors D_3 and D_4 count the numbers of photons passing the two respective final routes 3 and 4.*

The results obtained from the two versions seem incompatible: the open version demonstrates that a photon can travel only one of two available routes; the closed version implies that it should travel both routes simultaneously.

Wheeler's delayed-choice experiment was supposed to answer the question: what happens in reality? Does a photon travel only one of the two available routes or both? His idea was to choose which of two versions was to be performed by introducing or removing the second beam splitter, after the photon had already encountered the first beam splitter. In Wheeler's words, this will have 'an unavoidable effect on what we have a right to say about the already past history of that photon'.

Subtle delayed-choice experiments have been realized on photons,[12] and their results may seem strange to anyone who does not distinguish between concrete and abstract objects, but one who does can easily predict them. In fact, the outcome does not depend on whether the choice of the version is made before or after the photon has encountered the first beam splitter. This means, if we recall that experiments are carried out on concrete photons, that 'interference' occurs while each concrete photon travels just one path. This *fact* decisively disproves the abovementioned common belief that in the case of interference each photon travels both routes simultaneously and interferes with itself. And that is the *experimental* answer to Wheeler's question.

Thus the delayed-choice experiments also prove unambiguously that concrete photons have corpuscular properties only. But what then is the nature of the interference-like effects observed in those experiments? To answer this question we should recall that those effects are revealed through the dependence of the ratio of the numbers of concrete photons passing the two final routes on the performance of the closed version. This indicates that interference is of an essentially statistical nature, just as in the case of particle diffraction (see preceding sections); the experimental statistical data obtained on concrete photons conform to the probability distributions related to the corresponding abstract photons. Differences in the performance of the closed version result in different experimental statistical distributions, corresponding to different probability distributions, and that is the qualitative explanation of the 'interference' in the case under consideration.

Its quantitative explanation involves the use of certain equations (and we are not introducing equations). Nevertheless, some hints concerning the problem and the explanation of the primary result of the delayed-choice experiments can be provided without any equations.

Since the statistical data in the delayed-choice experiments are registered by the detectors placed on the final routes *behind* the appropriate devices, the processes taking place within the devices themselves are of no immediate interest to us: the devices can be regarded as merely black boxes. Of interest to us are only the parts of the open and closed systems closely adjacent to the corresponding detectors, and photons can be regarded as free particles there. This fact dramatically facilitates the treatment of the problem and allows one to deal with the general case valid for any particles.

The delayed-choice experiments can now be represented schematically as follows. Let two interswitchable devices I and II correspond to the above open and closed versions, respectively (see Figure 17). Device I emits concrete free particles of one kind in two different directions 1 and 2 *at random*. The number of particles emitted for a certain time interval in each direction is recorded by the corresponding counters C_1 and C_2. If the time interval is large enough, the ratios of those numbers to their sum give the measured (approximate) values of the corresponding probabilities. The concrete free particles recorded belong to the corresponding subsets A_1 and A_2 of the whole set of particles emitted by device I, whereas the probabilities are related to the corresponding abstract free particles. If the emission is so weak that there is a long enough time interval between any two consecutive counts, the records made by the two counters never coincide in time, and it is precisely this fact which has been found when performing the 'open' version on photons.

Device II, whose use results in some interference-like effects, emits concrete free particles of the same kind in two different directions 3 and 4 *at random*, these particles belonging to the corresponding subsets A_3 and A_4 of the whole set of concrete particles emitted by the device. The relative

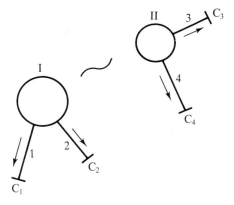

Figure 17. *Two interswitchable devices (I, II) correspond to open and closed versions, respectively. Device I emits concrete free particles of one kind in two different directions 1 and 2 at random, which are recorded by the respective counters C_1 and C_2. Device II emits concrete free particles of the same kind in two different directions 3 and 4 at random, which are recorded by the counters C_3 and C_4, respectively. Either device can be switched on by an experimenter at any time.*

numbers of the concrete particles recorded by the corresponding counters C_3 and C_4 for a long enough time interval are the measured (approximate) values of the corresponding probabilities related to the abstract free particles for device II.

In performing the experiment, one can arbitrarily switch either device on at any time, the results obtained obviously being determined by which one of them was switched on at the very last moment when the recorded concrete particles were emitted. This explains the primary result of the delayed-choice experiment, namely, that delaying one's choice of the experimentation mode has no effect on the outcome—again, the outcome is determined merely by the last-moment-choice, not by its delay.

Let us now look at the interference-like effects revealed when using device II. If subset A_3 in Wheeler's case consists of concrete photons that either passed through or were reflected at both beam splitters (see Figure 16), then subset A_4 consists of concrete photons that either passed through the first beam splitter and were reflected at the second one or vice versa. The interference-like effects occuring on either final route can then be treated as if the corresponding concrete photons have been emitted by two different sources. This allows one to outline the quantitative theory of 'particle interference' for the general case of any particles.

Suppose there are two sources, a and b, emitting *at random* concrete free particles of one kind (with equal energies and, hence, equal momentum magnitudes). They are elements of two different sets S_a and S_b, respectively.

The question that arises now is: what is the probability distribution of the free particle coordinates in this case?

To answer it, one should find the solution to the Schrödinger equation for this particular case, which describes the state of the abstract free particle corresponding to the whole set ($S_a + S_b$) of concrete particles emitted by the two sources. This solution has been shown[10] to be a wavefunction which is the sum of the wavefunctions corresponding to each of the sources a and b separately. Therefore, in view of the fact that probability distribution is proportional to the square of the magnitude of the wavefunction (see Section 6.1), the probability distribution in the case under consideration consists of three terms, one corresponding to source a alone, another corresponding to source b alone, and the third (so-called *cross term*) corresponding to both sources jointly. It is the cross term which is responsible for 'interference', that is, which makes the probability distribution in the case of two sources differ from a mere sum of the probability distributions corresponding to each of the sources separately.

Simple calculations show[10] that in Wheeler's particular case the predicted probability distribution is of exactly the same form that has been confirmed by the experiments.

One may wonder: how can a particle know which spot it should be in, and which spot must be avoided? The answer is simple. A particle 'knows' nothing—the concept of knowledge has nothing to do with the particle's behaviour. It is the connection between probability theory and experimental statistics which predetermines that the experimental statistical distributions *must* be in line with the calculated probabilistic predictions, provided that the latter are correct. Therefore, if the probability for a particle to be in some location is *exactly* zero, no concrete particle can get there. If the probability is not zero, some concrete particles *must* get there, but which ones of them is not predetermined either.

6.4 A brief discussion of the Heisenberg uncertainty principle

The conventional way of treating the double-slit Gedanken experiment leads, in particular, to the conclusion that a particle cannot move in a definite path (see Section 6.2). This is equivalent to the conclusion that it cannot have simultaneously definite coordinates and momentum. Why so? Because if a particle had definite coordinates (position) and momentum (and, hence, a definite velocity) at one definite moment, then at the next moment these would also be definite, viz., it would move in a definite path. Hence, both the coordinates and momenta of a particle are uncertain. But how uncertain: completely, or to some extent only?

To answer this question, one must first introduce a quantitative measure of an uncertainty. Heisenberg did this by considering some imaginary measurements of the coordinates and momentum of a particle. He regarded just the inaccuracies in these measurements as the respective uncertainties, and arrived at the conclusion that their product can never be less than the Planck constant h. (The constants h and were first mentioned in Section 5.2.) That was the first formulation of the famous Heisenberg uncertainty principle. The meaning of this principle is usually set forth as follows. It is impossible to measure simultaneously the coordinates and momentum of a particle with absolute accuracy—the higher is the accuracy of measuring one quantity, the lower is that of its counterpart. At the same time, it is usually argued that this impossibility has nothing to do with the technical accuracy of measurements: it is determined just by the absence of simultaneous values of coordinates and momenta for a particle, in principle.

At about the same time, the Heisenberg uncertainty principle had been generalized and derived theoretically, proceeding from considerations of quantum mechanics. Based on the probabilistic connection of the wave-function involved, the uncertainty of a physical quantity was defined in terms of probability theory, namely, as the standard deviation of the quantity in question, which is a non-negative quantity characterizing the mean deviation of the quantity from its average in the state under consideration. The derivation led to the conclusion that the product of the uncertainties of the generalized coordinates and momentum belonging to one coupled pair (see Section 5.4) cannot be less than /2.

The Heisenberg uncertainty principle usually is considered as valid without exception, and to be one of the most important features of conventional quantum mechanics. Some physicists even believe that this principle should be taken as the primary premise of quantum mechanics. The principle and especially its meaning have been discussed many times (in papers and at conferences), since there has always been confusion as to what this principle is all about. Most physicists, especially experimentalists, believe that it imposes restrictions on the accuracy of the measurement of coupled quantities. Some understand that it is related to the probability distributions of the values of those quantities, but usually without a full comprehension of what it means.

However, the results of EPR's Gedanken-experiment seemingly contradict this principle (see Section 3.1). The example of a free particle is also a counter-example which shows that this principle is not always valid. Indeed, the momentum of a free particle is constant. Hence, its uncertainty is precisely zero. The coordinates of a free particle are uncertain, but no matter how big the uncertainty, the product of both uncertainties remains zero too. Both examples show that the Heisenberg uncertainty principle needs some clarification, and we will now provide it.

Some readers, I believe, are already able to clear up the issue to some extent by themselves. Indeed, it suffices for them to recall that any experiment is carried out on concrete objects, whereas the principle in question belongs to quantum theory which deals with abstract objects. This means that the Heisenberg uncertainty principle has really nothing to do with measurements, for the latter necessarily involve some experimental procedures. As to Heisenberg's original experimental proof of the principle and other similar ones, a closer examination shows that they are ill-founded.[4]

Before discussing the theoretical proof of the principle as related to the product of the standard deviations of two mutually coupled quantities, let us note the following. The concept of standard deviations is related to any random variables (see Section 5.2), regardless of its origin. It is always a non-negative quantity by its very definition in probability theory. Therefore, the product of the standard deviations of *any* two random variables *must* be non-negative too, and this should be true also for any physical quantities which are random variables, by the very definition of this term in probabilistics (see Section 5.2). The example of a free particle is thus fully in line with this requirement.

Now, the theoretical proof of the Heisenberg uncertainty principle contends that the product in question can never be less than /2 (instead of zero). If its proof were valid for any possible case, then the particular case of a free particle would create a paradox. It has been shown, however, that the proof is valid under certain conditions only, which fail in the case of a free particle in particular, and in many others. An uncertainty principle that is always valid, without any exceptions, has been derived,[13,4] of which the Heisenberg uncertainty principle is a particular case. The new uncertainty principle yields correct results for the free particle, in particular, as well as for all the other cases which disagree with the Heisenberg uncertainty principle.

There is one more conventional prejudice in discussing the uncertainty principle. The principle presents an inequality, for it asserts that the product of two coupled standard deviations can never be less than /2. Many physicists replace the inequality by equality and contend that the product equals /2. From this they derive a conclusion that the smaller is the standard deviation of one quantity, the larger is that of its counterpart, and connect this result with the inaccuracies of the measurement of the quantities. That is, however, an altogether wrong conclusion, based on the mistaken replacement of an inequality by an equality. To illustrate this, it suffices to consider the case of a linear harmonic oscillator. A straightforward calculation of the standard deviations of its coordinates and momentum shows that they are equal (in proper units). This means that the larger is the standard deviation of one quantity, the larger is that of its counterpart, as opposed to what is usually believed.

Having considered some important quantum-related issues in the sequence in which they appeared, it is time now to return to the Bell's theorem related issues (see Sections 3.1–3.5).

6.5 Is the world real?

EPR's and Bohm's paradoxes, as well as some controversial experiments undertaken to check Bell's inequalities, have been considered in Sections 3.1–3.5 just as more or less simple examples of the application of the key principle which demands strict distinction between concrete objects and abstract objects (see Section 2.5). Bell's inequalities were mentioned only in passing there. Meanwhile, they are permanently attracting the attention of physicists, both theoreticians and experimentalists, and philosophers, and have done so for more than 25 years. Why is this? It is because they seemingly challenge the compatibility of quantum physics with reality. The challenge is presented in *Bell's theorem*,[14] which contends that a supposition of reality underlying quantum physics leads to inequalities (*Bell's inequalities*) which contradict quantum physics. Since quantum physics is in good agreement with experiments, whereas Bell's inequalities fail to comply with them, the conclusion one usually draws from this fact is that quantum physics cannot be based on assumptions involving the reality of the world. More exactly, Bell's theorem concerns so-called 'local realistic theories', that is, ones which proceed from both the reality of the physical world and the principle of 'locality'. The word 'locality' is used to express the conventional conviction, that no material thing can move at a rate exceeding the ultimate rate c of special relativity. In contrast to 'locality' is the concept of 'action at a distance', which means that something that happens to one physical system immediately affects another physical system, no matter how great is the distance between the two systems.

Thus, the experimental confirmation of the correctness of quantum predictions, not of Bell's inequalities, should mean that either the idea of the reality of the physical world is wrong, or there should be a kind of 'action at a distance'. If either inference were correct, that is, if Bell's theorem were right, it would indeed be a terrible blow to the realistic point of view in physics. However, as we shall see, the situation is not at all hopeless.

One might believe, for instance, that something is wrong with Bell's theorem itself. But it is apparently strictly proven mathematically. Hence, the only possibility which remains, if one wants to save the idea of the reality of the world, is to try to find out what is possibly wrong with the premises on which the proof of the theorem is based. To that end we should examine Bell's theorem as thoroughly as we can (but without any equations, as was promised).

The original Bell's theorem is related to Bohm's paradox, which concerns a pair of spatially separated spin-$\frac{1}{2}$ particles (1 and 2) with total spin zero (see Section 3.3). The spin component of each of the particles along any axis chosen can be measured by an appropriate device. Suppose there are three axes a, b, and c, which are the same for the both ends of the gap between the particles (see Figure 18). It is impossible to measure

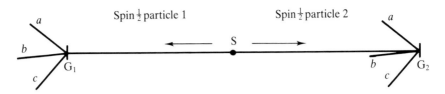

Figure 18. *Two spin-$\frac{1}{2}$ particles, 1 and 2, belonging to one pair with total spin zero, are emitted from the source S in opposite directions. At both ends of the setup there are special devices, G_1 (for particles 1) and G_2 (for particles 2), that can measure the spin components of each particle along each of the three different axes a, b, and c, which are common for both particles.*

the spin components of one particle along different axes simultaneously. Therefore, such an experiment can be performed only with a large number of pairs of particles 1 and 2 of one kind.

Let the spin components of particles 1 be measured along the *a* and *b* axes, and those of particles 2 along the *b* and *c* axes. Each measurement yields an accidental result which cannot be predetermined. Therefore, we are actually dealing with a statistical experiment. The accumulated statistical data can be used for calculating various arithmetic averages. Bell was interested in the theoretical calculation of the expected arithmetical averages of the following products of the spin components of the particles: the product of spin components of particle 1 along the *a* axis and of particle 2 along the *b* axis; the product of spin components of particle 1 along the *a* axis and of particle 2 along the *c* axis; and the product of spin components of particle 1 along the *b* axis and of particle 2 along the *c* axis.

The counterpart in probability theory of the statistical arithmetic average is the mathematical expectation (see Section 5.2). Therefore, Bell had to calculate the mathematical expectations of the three aforementioned products. He demonstrated that, under assumptions he had made (which he believed to satisfy the conditions of 'reality' and 'locality'), those three mathematical expectations were tied together by an inequality (*Bell's inequality*) which contradicted the predictions of quantum mechanics. The mathematics in Bell's reasoning seems irrefutable, as I mentioned before. However, what about the assumptions he utilized to get the inequality? Are they truly 'realistic'? We shall examine that now.

His reasoning can be set forth as follows. Considering Bohm's Gedanken experiment (see Section 3.3), one may infer that the result of measuring any chosen spin component of particle 2 must be predetermined by previously measuring the same spin component of particle 1. Since the wavefunction describing the state of the pair does not determine the result of an individual measurement, this predetermination implies that there should be a more complete specification of the state.

Suppose there are various specific states in which the result of an individual measurement is determinate, each state having a definite probability of existence. Suppose, too, that the result of measuring in each state the spin component of either of particles 1 and 2 along the three axes *a*, *b*, and *c* is determined by the choice of the axis and by the specific state only. It is these two assumptions which define in Bell's theorem the realistic states satisfying the requirement of locality (since no action at a distance is supposed). It is then the utilization of these 'realistic states' which leads to Bell's inequalities.

However, a definition of realistic states must take into account the *real* properties of the physical systems involved, otherwise the definition just cannot work. It is precisely this which happened in the present case. The assumptions in Bell's theorem actually suppose that there are common states for different axes chosen in which the spin of either particle has definite components along different axes simultaneously. But, that is a wrong supposition, as was explained in Section 3.3. No common states may exist in reality for different axes along which the spin components of a particle are considered, for the simple reason that different axes do not exist simultaneously—they cannot be chosen simultaneously! If, for instance, in some state all the concrete particles 2 have the definite spin component 1 along the *b* axis, when this axis is chosen, then each of them also has a definite spin component along the *c* axis, when this is chosen. But which of the values ± 1 it has is not predetermined.

In short, the assumptions of Bell's theorem contradict the real properties of concrete spin-$\frac{1}{2}$ particles. No wonder that Bell's inequalities fail to comply with real experiments!

At the same time, as has been shown[4, 15] the calculation of the mathematical expectations in question, which takes into account the real properties of concrete spin-$\frac{1}{2}$ particles, leads to results that are fully in line with experiment and with the predictions of quantum physics (since the latter is merely a particular domain of probabilistic physics). Here are the outlines of the calculation.

It concerns each pair of the axes *a*, *b*, and *c*, separately. For each pair of axes there are only four real states, in accordance with the following possible combinations of the spin components along these axes: $(1, 1)$; $(1, -1)$; $(-1, 1)$; $(-1, -1)$. The probability is $\frac{1}{2}$ for the state in which the spin component of particle 1 along the first axis is 1. The probability that in *this* state the spin component of particle 1 along the other axis equals 1 or -1 can be obtained from relatively simple considerations of probabilistics, no specific quantum-mechanical terminology being involved. The same is true for the state in which the first spin component is -1. The required products of the spin components in these states are, respectively: $1, -1, -1, 1$. The mathematical expectation of the product is merely the arithmetic average of the latter over all the four states.

We have considered here only the original Bell's theorem. There are many other versions of it, but they all suffer from one and the same failure—the lack of distinction between concrete objects and abstract objects and the failure to take into account the real properties of the concrete particles involved. The version related to the famous Aspect's experiments deserves special consideration for a very curious reason—it had actually been disproved long before it was derived! Here is how and why.

This version of Bell's inequalities involves the mathematical expectation of the product of the polarizations (*p*-values) along the respective *p*-lines of the two photons belonging to one emitted pair (in Aspect's terms, the coefficient of the correlation of polarization; see Sections 3.4 and 3.5). Two *p*-lines (a and a′) refer to one photon, and the other two (b and b′) to its counterpart in the pair. The calculation, based on allegedly 'realistic' assumptions like those used in the spin-$\frac{1}{2}$ particle case, leads to the conclusion that the mathematical expectations of the products in question related to the four combinations of *p*-lines (a, b), (a, b′), (a′, b), (a′, b′) , are tied together by an inequality (*Bell's inequality*) which contradicts experiment and quantum mechanical predictions (the latter being in good agreement with experiment). The reason for the experimental failure of Bell's inequalities in this case is the same as in the previously considered case, namely, lack of distinction between concrete and abstract objects and failure to take into account the real properties of the concrete particles involved.

Bell's inequalities for Aspect's case were doomed to failure beforehand because if they were correct the Malus law would be wrong (those inequalities are incompatible with the Malus law). However, the Malus law was both well established experimentally and obtained theoretically from simple considerations of classical electrodynamics. Therefore, any theoretical results which would disprove the Malus law must be notably wrong.

What does the Malus law have to do with Aspect's experiments? The answer is simple. To predict the results of those experiments it suffices to make use of the Malus law with due regard for the 'polarization conservation law' (see Section 3.5). Neither quantum mechanics, nor any other probabilistic consideration is needed.

Indeed, the Malus law concerns the dependence of the intensity of light passing through an analyser on the angle α between the polarization of the light and the *p*-line of the analyser (the intensity is proportional to $\cos^2\alpha$; see Section 3.4). In terms of photons, however, the intensity of the light passing through the analyser is proportional to the probability that the polarization of a photon is parallel to the *p*-line of the latter ($p = 1$ for the photon). Hence, if a photon is polarized parallel to some *p*-line chosen ($p = 1$ with respect to this *p*-line), the probability that it will have $p = 1$ with respect to another *p*-line is $\cos^2\alpha$, where α is the angle between the two *p*-lines. The second photon of the pair, due to the 'polarization conservation law', has the same *p*-value with the same probability that the first one has with respect to each of the *p*-lines in question.

The calculation of the mathematical expectations needed is very simple now. There are only four states for the abstract pair of photons under examination, in accordance with the four possible combinations of the polarizations of the photons (*p*-values) along any two *p*-lines: (1, 1), (1, − 1), (− 1, 1), (− 1, − 1). The probability of the state, in which one photon has $p = 1$ along one *p*-line is 1/2. The probability that it has $p = - 1$ along that *p*-line is the same. The products of the *p*-values of the photons in these states are, respectively, 1, − 1, − 1, 1. Given the probabilities of the products (from the Malus law), the calculation is simple arithmetic. The result is fully in line with Aspect's experiments, as it should be, for the latter can be regarded as merely the verification of the Malus law (and the 'law of conservation of polarization').

Incidentally, the probabilities needed can be obtained, too, from the same considerations of probabilistics as in the spin-$\frac{1}{2}$ particles case.[15] They are the same as those given by the Malus law, which means that those probabilistic considerations can be regarded as the theoretical proof of the Malus law in terms of photons.

The conventional opinion that Bell's inequalities are based on 'local realism', whereas quantum mechanics is alien to it, is due to a misunderstanding. The actual situation is quite the opposite. It is precisely quantum physics which deals with reality, as experiment shows, whereas Bell's inequalities fail the experimental test just because they result from unrealistic assumptions. The misunderstanding is due partly to the wrong construction of conventional quantum mechanics, which gives the impression that it has nothing to do with reality. But already there is quantum physics, meant as a whole science, which leads to basically the same mathematics as that in conventional quantum physics, but appears as a particular domain of probailistic physics based on fully realistic premises (see Section 5.3). Hence, probabilistic quantum physics as a whole science is fully compatible with reality, and Bell's challenge is a total failure.

There is one more point which invalidates arguments like Bell's. As was mentioned in Section 3.1, Bell's-inequalities-related experiments are mistakenly identified with particular versions of EPR's Gedanken-experiment. This leads to the actual substitution of *determinism* (characteristic of EPR's case) for *realism*. Thus the only thing Bell's-like-arguments can prove, as a matter of fact, is that quantum physics is incompatible with *determinism* (not with *realism*). But this is quite obvious and needs no special proof, for quantum physics is closely connected with *chance*, whereas determinism is inconsistent with it.

We are already approaching the end of the book. To be sure, there are other quantum related problems not covered, but *this* book is not intended to consider all the problems nowadays under discussion. There is, however, one which I want to present to the reader right now. We were referring to a photon, in the course of the exposition, as a regular particle. Also, in the experiments carried out on particles, just photons were used as a rule. This means that there should exist *concrete* photons, as in the case of any other particles used in experiments. But what right do we have to speak of photons in the same terms as we do when dealing with regular particles? It is this issue which will be discussed next.

6.6 The mystery of the photon

In Section 6.1, the question of what is a particle was discussed in connection with the wave–corpuscle duality. It was demonstrated that *concrete* particles have merely corpuscular properties, whereas wave properties are related to *abstract* particles and connected with (theoretical) probability distributions of the particle coordinates. The results obtained there are valid for any particles, regardless of their nature and the inherent properties they possess, such as, for instance, mass, charge, etc. Hence, they are valid for photons as well. This means that a concrete photon also has corpuscular properties only, and should behave in mechanical respects like any other concrete particle.

What is the problem with a concrete photon, then? Why does the question of what is a photon arise at all?

The answer is: it arises because of the following supposed strange behaviour of a photon. A regular particle can move at any speed less than the ultimate rate c of special relativity, which is believed to be that of light (or electromagnetic waves) in a vacuum. It cannot move exactly at c, for its energy, according to special relativity, would become infinite, which is impossible. But its velocity can be zero, in particular, which means that a regular particle can stop (be at rest), without changing its identity. That is not true for a photon, as usually believed. A photon is supposed to move exactly at c, for it is considered a 'quantum' of light. Therefore, it can move at no other rate; in particular, it cannot stop (be at rest) without changing its identity. If a photon stops, it simply disappears. When a regular particle stops, it still has a definite nonzero mass called its 'rest mass'. The rest mass of a photon should thus be exactly zero, according to conventional belief, otherwise it could not move exactly at c; instead it would be able to move, like any regular particle, at any rate less than c.

Consequently, the question of what is a photon concerns, first of all, its rest mass. This and other related questions bring up a more general one: are there any concrete particles in reality whose rest mass is zero?

A correct answer to this question is very important, because it concerns not only photons, but also other particles, in particular neutrinos whose rest mass is usually believed to be zero as well. Then, if the answer to this question is in the negative, that is, if concrete particles with zero rest mass do not exist in reality, a concrete photon cannot have zero rest mass either—it should have some definite, perhaps very small, but nonetheless nonzero rest mass. This inference would lead, however, to far-reaching consequences. A concrete photon could no longer be regarded as an outcome of the electromagnetic field, as usually interpreted. Instead, the electromagnetic field, a free one at least, would be due to the real existence of the concrete photons of nonzero rest mass. This development would demand the reconstruction of classical electrodynamics, which would require a new, more profound insight into many matters now regarded as well known.

What, then, is the answer to the above question? Can a concrete particle have zero rest mass or can it not?

A careful examination of this question demonstrates that it cannot.[4,16] This means, in particular, that a concrete photon also has nonzero rest mass. Hence, a concrete photon can move, like any other concrete particle, at any rate *less* than c, and even be at rest. When at rest or moving at low speed (much lower than c), it can no longer be regarded as a 'quantum' of light, and the name 'photon' becomes somewhat inappropriate. For this reason and because of its relation to the electromagnetic field, it makes sense to call this subatomic particle which becomes a photon when moving at rates near c, the *emon*. The first two letters in the name reflect the particle's electromagnetic association.

Incidentally, Louis de Broglie was the first to contend that radiation had to be considered as divided into atoms of light of very small mass, and that electromagnetic theory required revision.

The photon–emon connection will be examined now in more detail. A concrete emon can move at any rate less than c. Every concrete emon moving at a velocity very close to c is a concrete photon. The higher is the speed of a concrete emon (the closer it is to c), the higher is the energy of the corresponding photon. But, according to Einstein's relation (mentioned in Section 6.1), the 'frequency' of a photon is proportional to its energy. Therefore, the closer is the speed of a concrete emon to c, the higher is the 'frequency' of the corresponding photon. It turns out that the entire interval of frequencies corresponding to all the electromagnetic waves now available is due to the motion of concrete emons within a very narrow range of speeds near c. The reader should be reminded, by the way, that the concept of frequency is alien to a *concrete* photon (see Section 6.1)—it is related to the wavefunction describing the state of the corresponding *abstract* photon conforming to the entire set of concrete photons with the same value of energy. Keeping this in mind, we shall, nonetheless, use the term 'frequency of a photon' for convenience.

There is an exact relation between the velocity of a concrete emon and the frequency of the respective photon, and it involves the rest mass of the emon. The task of finding the exact value of that rest mass is, however, extremely hard if not altogether impossible, for the time being, at least, for the mass is extremely small. Proceeding from some established experimental data, one can show that it is smaller than the rest mass of the hydrogen atom divided by 10^{24} (that is, by 1 with 24 zeros). It is quite possible that it is even much smaller than that estimate, and minute mass is impossible to measure by any direct method. Also, the accuracy of measurement of the velocity of light does not allow one to assert that it is constant regardless of the frequency of the light. Thus there is no experimental proof that the velocity of light is a constant—the ultimate velocity c of special relativity— and that the rest mass of a photon (that is, of the emon) is zero.

Anyhow, the rest mass of the emon is inconceivably small. Why should we, then, bother about it? Why not simply ignore it entirely?

Certainly, the fact that the emon has nonzero rest mass changes nothing in the conventional applications of electrodynamics. But, as was mentioned before, the importance of this fact can hardly be over-estimated. First of all, the velocity of light is no longer a constant—it is different for different wavelengths and never coincides with c, even though it can approach c very closely. The relative difference in the velocity from c (the ratio of the difference to c) is less than $1/10^{10}$ for very long (radio) waves and $1/10^{40}$ for the hardest gamma rays, provided that the above estimate of the emon rest mass is correct. Since the actual emon rest mass could be much smaller, the emon velocities corresponding to the indicated electromagnetic wavelengths could be even closer to c.

Next, photons are no longer the outcome of the electromagnetic field. On the contrary, the electromagnetic field is the outcome of the existence of the emon. The reader may recall, by the way, that the concept of an electromagnetic field had been introduced originally as an auxiliary mathematical quantity facilitating calculations. Only much later did it become regarded as an independently existing entity obeying, in particular, the famous Maxwell equations. Recognition of the existence of the emon leads us to reconsider the electromagnetic field as an auxiliary mathematical quantity. The new situation demands, however, a new approach to the treatment of *all* electromagnetic phenomena. The link between the electromagnetic field and emons should be understood, and the Maxwell equations have to be derived from the emon considerations, as the first step in the new construction of classical electrodynamics.

This task has already been accomplished.[17] It transpires that a concrete emon at rest possesses a specific *electric* property *only*. Its specific magnetic property appears only when the concrete emon moves at velocities very close to c, that is, when it becomes a photon, due to the requirements of special relativity. An interesting fact should be noted, too: that the free electromagnetic field depends on the emon rest mass in such a way that, if the latter were zero, so would be the former. This corroborates the assertion that the emon should have a nonzero rest mass.

After obtaining the Maxwell equations, the task of the construction of classical electrodynamics, proceeding from the new approach, is not yet complete, and there is a long way ahead. One should explore and understand, first of all, the nature of the interaction between emons and charged particles, in order to obtain the dynamic laws, and so on. There are many other interesting problems as, for instance, the part the slow emons may play in biochemical and neurological processes; or perhaps they are responsible, at least partly, for the so-called 'missing mass' in the universe, etc.

One cannot foresee what kind of new developments may occur in the future in connection with that new approach, based on the recognition of the real existence of the emon. Nevertheless, it is my belief that we are arriving at the threshold of a new era in physics and related sciences.

Epilogue

It has been a long journey from Schrödinger's cat to the new developments connected with the existence of the emon. Along the way, some readers might have experienced difficulties in digesting new concepts, alien ideas, and strange approaches. Some aspects could, of course, have benefitted from more extensive contemplation, but I hope that all readers have successfully overcome any 'stumbling blocks' arising out of the brevity of the discourse, and gained a reasonably clear understanding of our material. Nonetheless, it may, I believe, be helpful now to review the prime logical line of the book (and some of its ramifications) with the emphasis on its most important points.

We started with the exposition of Schrödinger's cat paradox caused by the conventional quantum mechanical treatment of Schrödinger's Gedankenexperiment (Section 1.2). This brought up the general problem of paradoxes, the brief examination of which led to the conclusion that the settlement of the cat paradox and some others, including the chicken–egg paradox, requires a new way of understanding things—a new principle of thinking (Section 1.3). A careful investigation of Schrödinger's cat experiment made us suspect that the related paradox is due to some flaw in the interpretation of the conventional quantum mechanical formalism (Section 1.4). Since the key word in that interpretation is 'probability', we turned to tackling its meaning (Section 2.1).

It was at this point that we were brought to the necessity of changing our usual way of looking at and speaking of things. It proves necessary to understand clearly the difference between things that really exist and with which we deal in our everyday life and in any experiments (hence, in the experimental part of a science), and things we merely talk about (dealt with in theoretical considerations and, hence, in the theoretical part of science), which do not exist in reality (Sections 2.2–2.4). It is the confusion of the former (called concrete objects) and the latter (called abstract objects) which, due mainly to the fact that the names of both generally coincide, leads not infrequently to paradoxes, such as the cat paradox and the chicken–egg paradox in particular. We thus come to the inevitability of a new way of looking at and describing things, namely, by keeping in mind the necessity for the strict distinction between concrete objects and abstract objects (Section 2.5).

That is the key principle whose importance can hardly be overestimated. It is a principle that should be taken into account in each and every science, including the humanistic ones, and perhaps even in our everyday life if required. This principle allows us, in particular, to realize finally that probability theory and experimental statistics constitute both the theoretical

91

and experimental parts of one unique science of probability, called 'probabilistics' (Section 5.1). The true realization of this fact, makes it possible, in turn, to identify the basic phenomenon of this science and to give an explicit general definition of probability, all the conventional rules of dealing with probability and related quantities follow from that definition. Experimental statistics deals with concrete objects, whereas probability theory is related to the corresponding abstract objects, and an important point to be emphasized is that the statistical frequency of a random event in a large series of random tests is the measured (hence, approximate) value of the corresponding probability.

Probabilistics provides at least two methods of problem solving: the classical method and the quantum approach. The latter involves specific mathematical techniques used, in particular, in conventional quantum mechanics, but belonging (I accentuate this again) to probabilistics itself (Section 5.2). The application of the classical method to physical *macrosystems* leads to classical statistical mechanics. This method is not good for *microsystems*—only the quantum approach is relevant for them, which leads directly to quantum physics (Section 5.3). *In total*, classical statistical mechanics and quantum physics are two interconnected domains of one unique science, probabilistic physics, which arises as a result of the application of probabilistics to physics in general.

Then came the basic equation of quantum mechanics, the famous Schrödinger equation, which results directly from the requirements of the reality of motion of concrete mechanical systems, which is reflected in the equality of the mathematical expectations of two different physical quantities, the Hamiltonian function and the energy (Section 5.4). We thus came to the point which can be regarded as the end of the principal logical line of this book. There are some ramifications, however.

1. The immediate application of the key principle easily settles the chicken–egg and Schrödinger's cat paradoxes, both having equally no bearing on any physical considerations (Section 2.5). This principle also settles, without difficulty, more complicated cases of paradoxes, namely, EPR's, Bohm's, and those related to Bell's inequalities; this time specific physical laws (corresponding conservation laws) are involved (Sections 3.1–3.5, 6.5).

2. The issue of wave–corpuscle duality is settled in a fully realistic way, and the origin of related wave phenomena, such as the diffraction and interference of particles, is clarified (Sections 6.1–6.3). Concrete particles have merely corpuscular properties, whereas the so-called 'wave properties' are related to the probability distributions referring to corresponding abstract particles. The wave phenomena are thus of an essentially statistical nature.

3. Examination of the related problem of the interrelation between the electromagnetic field and photons leads to the assertion of the existence of a subatomic particle, the emon, and to prospective new developments in physics and other sciences (Section 6.6).

4. The Heisenberg uncertainty principle is corrected, and its real meaning is explained (Section 6.4). The important point is that it has no bearing whatsoever on measurements.

5. The application of our key principle to social problems is an especial ramification of the main line of this book (Sections 4.1–4.6). It is included not only because it illustrates how the principle works, but also in view of the extreme significance of the matters discussed there, which are no less important than those related to physics and other natural sciences.

I hope that this quick summary may help the reader to hold in mind the book's content more clearly. Also, I should like to believe that some readers may be inspired by it to try to engage themselves in the exciting adventure of searching for truth—an adventure which can be called genuine science when and only when it proceeds from a clear understanding of reality.

* *
*

Feci quod potui, faciant meliora potentes
I have done what I could, let the one who can do it better

References

1. Mayants, L. S. 1973, *Foundations of Physics*, **3**, 413.
2. Schrodinger, E. *Naturwissenschaften*, **23**, 812.
3. Mayants, L. S. 1970, *Zapiski Erevanskogo Gosud. Universiteta*, **2**, 156.
4. Mayants, L. S. 1984, *The Enigma of Probability and Physics*, Dordrecht: Reidel.
5. Einstein, A. Podolsky, B. and Rosen, N. 1935, *Physical Review*, **47**, 777.
6. Bohm, D. 1951, *Quantum Theory*, Englewood Cliffs: Prentice Hall.
7. See, for instance, Aspect, A. 1983 Lingren, I., Rosen, A. and Svanberg, S. (eds), *Atomic Physics* vol. 8, p. 103, New York: Plenum
8. Shaw, G. B. 1938, Preface to *On The Rocks*, in *Collected Prefaces*, p. 163, Oldham's Press.
9. Mayants, L. S. 1977, *Foundations of Physics*, **7**, 3.
10. Mayants, L. 1989, *Annales de la Fondation Louis de Broglie*, **14**, 177.
11. Wheeler, J. A. 1978, in Marlow, A. R. (Ed), *Mathematical Foundations of Quantum Theory*, p. 9, New York: Academic Press.
12. Helmuth, T., Walther., H., Zajoonc, A. and Schleich, W. 1987, *Physical Review A*, **35**, 2532.
13. Mayants, L. S. 1974, *Foundations of Physics*, **4**, 335.
14. Bell, J. S. 1965, *Physics*, **1**, 195.
15. Mayants, L. 1991, *Physics Essays*, **4**, 178.
16. Mayants, L. S. 1981, *Foundations of Physics*, **11**, 577.
17. Mayants, L. 1989, *Physics Essays*, **2**, 223.

Glossary

Angular momentum: A physical quantity related to the rotational motion of a body; determines the rotational energy of the body.

Blackbox: A device whose inner structure and workings are unknown and do not need to be known.

Chord: A line segment which joins two points on a circle.

Classical mechanics: A domain of physics which deals with the mechanical systems obeying Newton's laws.

Conservation law: A law stating that under certain circumstances the value of a physical quantity for a particular physical system does not change with time.

Diffraction: A change of the wave pattern when a wave encounters obstacles (e.g., of waves on the surface of water).

Electric vector: See *Electromagnetic field*.

Electrodynamics: A domain of physics which deals with electromagnetic phenomena.

Electromagnetic field: A domain of space exhibiting varying electric and magnetic properties at each its point; these properties being vectors.

Electromagnetic wave: An electromagnetic field propagating as a wave, for instance, light.

Galileo's relativity: A relativity principle stating that the acceleration of a body does not depend on the choice of an inertial coordinate system.

Gamma rays: High frequency radiation emitted by some radioactive atoms.

Gedanken-experiment: An experiment which is meant to be performed mentally—not in reality.

Geiger (*Geiger counter*): A device which allows one to register an event that causes a discharge in the device.

Generalized coordinates: Appropriate quantities describing the spatial configuration of a mechanical system.

Generalized momenta: Appropriate quantities that complement the generalized coordinates in a complete description of a mechanical system.

Geometric optics: A domain of optics which deals with beams of light.

Hamiltonian equations: A set of specific equations involving generalized coordinates, generalized momenta, and time, and which are valid for mechanical systems of certain important types.

Hamiltonian function: A special function of generalized coordinates, generalized momenta, and time, and which has the dimension of energy.

Harmonic oscillator: A body vibrating with a constant frequency.

Inertial coordinate system: A coordinate system with respect to which a free body moves at a constant rate.

Interference: Change of a wave pattern as a result of wave overlap.

Linear algebra: A branch of mathematics which deals with linear operations on vectors.

Magnetic vector: See *Electromagnetic field*.

Matrix: A generally rectangular table filled in with numerical or algebraic quantities, regarded as one entity.

Momentum: A physical quantity characteristic of the translational motion of a body; determines the translational energy of the body.

Neutrino: A subatomic neutral particle of extremely small mass.

Operator: A mathematical quantity in linear algebra which transforms any vectors into generally different vectors.

Photo-effect: An emission of electrons by a metal surface as a result of illuminating it by a specific source of light.

Planck's constant: The constant that connects the energy of a photon with its frequency.

Random event: An event which is the result of chance alone.

Random test: A test performed in such a way that its result is entirely accidental.

Special relativity: Einstein's special relativity theory.

State vector: A mathematical vector describing a state of an abstract physical system.

Statistical experiment: An experiment that provides for fully accidental results.

Vector (in real physical space): A quantity having both a magnitude (length of the vector) and a direction.

Vector (in linear algebra): A set of numerical or algebraic quantities, regarded as one entity, which obey the same rules of addition and multiplication as do real space vectors (q.v.).

Wave optics: A domain of optics which deals with the wave properties of light.

Index

Abstract object, 11, 13–14, 18–22, 27, 31, 37–39, 41–42, 51–53, 56, 60–61, 75, 80–81, 84, 90–91
  concept of, 11, 14, 18, 37
  degree of abstraction, 19
  name, 19–20
  principal features, 20
  state of, 56–57, 59–60
Abstraction, 20
Accidental result, 10, 82
Achilles, 4
Action at a distance, 26, 32, 35, 81, 83
Adequate sets, 52–55
Analyser, 32–34, 84
Angular momentum, 28–30, 94
  conservation law, 30
  peculiarities, 30
Arago, 67
Arithmetic average, 58, 82
  statistical, 82
Aspect, A., 32
Aspect's experiment, 32–35, 52, 84–85
At random, 7, 10, 31, 52, 54, 56–57, 76–77
At rest, 86–88

Balls, 10–15, 51, 53–54, 57
  abstract, 11, 13, 51
  concrete, 11, 13, 15, 51
Beam, 2, 69
  concrete particles, 2, 69
  electrons, 2, 7
Beam splitter, 74, 75
Bell, J. S., 82
Bell's inequalities, 27, 40, 81, 83–85, 91
Bell's theorem, 32, 80–81
  premises, 81, 83

Bertrand, J., 9
Black box, 76, 94

Cat, 2, 6, 14
  abstract, 23
  concrete, 6, 23
  state of, 2, 3, 23
Cat experiment, 2, 3, 6, 7, 23
Chord, 9, 94
  concrete, 52, 54
Class, 13, 16–18, 44–46, 51, 53, 57
  definition, 16
  elements of, 16–17, 19
  empty, 17–18
Classical mechanical system, 55
Classical mechanics, 5, 24–25, 27, 29, 59, 62–63, 94
Classical method, 55, 60–61, 64, 91
Classical statistical considerations, 54, 60
Classical statistical mechanics, 5, 64, 91
Classical statistical physics, 38
Communication, 41
Communist regime, 39
Compton effect, 68
Computer, 7, 16, 57, 59
Concrete objects, 6, 11, 14–16, 18–20, 22, 26–27, 31, 37–39, 51–54, 56–58, 60–61, 75, 80–81, 84, 90–91
  classes, 16–20
  concept, 11, 14, 18, 37
  properties, 20, 27, 40
  set of, 19, 51–52
Conservation laws, 37, 91, 94
Coordinates, spatial, 63–64
Cross term, 78

99